DESIGN-BUILD FOR THE DESIGN PROFESSIONAL

2001 Supplement

DESIGN-BUILD FOR THE DESIGN PROFESSIONAL

2001 Supplement

G. William Quatman II, FAIA, Esq.

ASPEN LAW & BUSINESS
A Division of Aspen Publishers, Inc.
New York   Gaithersburg

Copyright © 2001 by Aspen Law & Business
A Division of Aspen Publishers, Inc.
A Wolters Kluwer Company
www.aspenpublishers.com

All rights reserved. No part of this publication may be reproduced or transmitted in any form or by any means, electronic or mechanical, including photocopying, recording, or any information storage and retrieval system, without permission in writing from the publisher. Requests for permission to make copies of any part of the work should be mailed to:

Permissions
Aspen Law & Business
1185 Avenue of the Americas
New York, NY 10036

Printed in the United States of America

1 2 3 4 5 6 7 8 9 0

Library of Congress Cataloging-in-Publication Data

Quatman, G. William, 1958-
  Design-build for the design professional / G. William Quatman
    p. cm.
  Includes bibliographical references and index.
  ISBN 0-7355-1727-4
  ISBN 0-7355-2183-2 (supplement)
    1. Construction contracts—United States. 2. Architectural contracts—United States. I. Title.

KF902 .Q38 2001
345.73′07869—dc21

00-063955

About Aspen Law & Business

Aspen Law & Business is a leading publisher of authoritative treatises, practice manuals, services, and journals for attorneys, corporate and bank directors, accountants, auditors, environmental compliance professionals, financial and tax advisors, and other business professionals. Our mission is to provide practical solution-based how-to information keyed to the latest original pronouncements, as well as the latest legislative, judicial, and regulatory developments.

We offer publications in the areas of accounting and auditing; antitrust; banking and finance; bankruptcy; business and commercial law; construction law; corporate law; criminal law; environmental compliance; government and administrative law; health law; insurance law; intellectual property; international law; legal practice and litigation; matrimonial and family law; pensions, benefits, and labor; real estate law; securities; and taxation.

Other Aspen Law & Business products treating construction law issues include:

Alternative Clauses to Standard Construction Contracts (Second Edition)
Architect and Engineer Liability: Claims Against Design Professionals (Second Edition)
Calculating Construction Damages (Second Edition)
Calculating Lost Labor Productivity in Construction Claims
California Construction Law (Sixteenth Edition)
Construction Bidding Law
Construction Change Order Claims
Construction Claims Deskbook: Management, Documentation, and Presentation of Claims
Construction Defect Claims and Litigation
Construction Delay Claims (Third Edition)
Construction Disputes: Representing the Contractor (Third Edition)
Construction Documentation (Third Edition)
Construction Industry Forms (Second Edition)
Construction Joint Ventures: Forms and Practice Guide
Construction Law Handbook
Construction Law Library on CD-ROM
Construction Law Update (Annual)
Construction Litigation: Practice Guide with Forms
Construction Litigation: Representing the Owner (Second Edition)
Construction Management: Law and Practice
Construction Scheduling: Preparation, Liability and Claims
Construction Subcontracting Manual: Practice Guide with Forms
Construction Surety and Bonding Handbook
Design-Build Contracting Claims
Design-Build Contracting Formbook
Design-Build Contracting Handbook (Second Edition)

Differing Site Condition Claims
Fifty State Construction Lien and Bond Law (Second Edition)
Florida Construction Law
Handling Construction Defect Claims: Western States (Third Edition)
Legal Guide to AIA Documents (Fourth Edition)
Practical Guide to Construction Contract Surety Claims
Proving and Pricing Construction Claims (Third Edition)
State-by-State Guide to Architect, Engineer, and Contractor Licensing
Sweet on Construction Industry Contracts: Major AIA Documents (Fourth Edition)

ASPEN LAW & BUSINESS
A Division of Aspen Publishers, Inc.
A Wolters Kluwer Company
www.aspenpublishers.com

CONTENTS

Sections listed below appear only in the supplement and not in the main volume.

Chapter 1	**The Growth, Opportunities, and Appeal of Design-Build**	**1**
§ 1.15	The International Design-Build Market	1
	[D] Canada	1
Chapter 2	**The History of Design-Build**	**3**
§ 2.08	Canadian Design-Build Organizations	3
	[A] The Canadian Construction Association (CAA)	3
	[B] Canadian Design-Build Institute (CDBI)	3
Chapter 4	**Variations in Design-Build**	**5**
Chapter 5	**Designer as Subcontractor to a Contractor**	**7**
Chapter 6	**Bridging**	**9**
§ 6.01	What Is Bridging?	9
	[D] Bridging in Canada	9
Chapter 7	**Teaming for Design-Build Projects**	**11**
Chapter 9	**Business Issues for Designers in Design-Build**	**13**
Chapter 12	**Contractor and Professional Licensing Laws**	**15**
§ 12.19	Canadian Licensing Law Requirements	17
	[A] Corporations	18
	[B] Partnerships	19
	[C] General Requirements	19
	[D] Design-Led Design-Build	20
Chapter 13	**Insurance and Bonding for Design-Build**	**23**
Chapter 14	**Public Sector Design-Build**	**25**
§ 14.15	Payment of Stipends	27
	[G] Payment of Stipends in Canada	27

Chapter 15   **Design-Build and the Federal Government**   **29**

Chapter 16   **Ownership of Design Proposals, Technical Submissions, and
                     Copyrights**   **31**
§ 16.13   Copyright Issues in Canada   32

Chapter 17   **Design-Build Project Types and Project Studies**   **33**
§ 17.14   Hospitals   37

Chapter 18   **Design-Build Contracts**   **39**
§ 18.17   Canadian Contract Forms   39
                     [A]   Canadian Construction Documents Committee (CCDC)
                           Forms   39
                     [B]   The Association of Consulting Engineers of Canada's
                           Opposition   41

Chapter 20   **Project Management**   **43**

Chapter 21   **Design-Build Risks and Liabilities**   **45**

Chapter 22   **Job Site Safety**   **47**

Chapter 23   **When Design Goes Bad: Lessons Learned for Litigation**   **49**
§ 23.12   Action for Fraud Based on Misrepresentation   54
                     [C]   Action for Fraud Based on Scope Change by Owner   54

Chapter 24   **Ethical and Criminal Issues in Design-Build and
                     Construction**   **57**
§ 24.01   Conflicts of Interest   57
                     [E]   Adverse Effects of Conflicts   57
§ 24.03   Making Decisions on Pay Applications, Completion Dates,
                     or Claims   58
                     [C]   The Canadian Perspective   58
§ 24.12   Governing Laws   59
                     [I]   Canadian Price Lists and Tariffs   59
§ 24.16   Ethical Issues in Canada   60
                     [A]   Conflicts of Interest   61
                     [B]   Providing Free Services   62
                     [C]   Sharing Stipends   63
                     [D]   Compensation from More than One Client   63
                     [E]   Compensation Derived from Construction Profits   63
                     [F]   Design Competitions Restricted   64
                     [G]   Competition on Fees   64
§ 24.17   Interference with Contract   65

Appendix B1   **Supplementary Conditions to AIA Document No. B901 (1996
                     Edition); Standard Form of Agreement Between Design-
                     Builder and Architect**   **67**

CONTENTS

Appendix D1   Electronic Data Agreement   73

Appendix E   State Design-Build Statutes (Including Attorney General Opinions)   75

Index   85

DESIGN-BUILD FOR THE DESIGN PROFESSIONAL

2001 Supplement

THE GROWTH, OPPORTUNITIES, AND APPEAL OF DESIGN-BUILD

§ 1.15   THE INTERNATIONAL DESIGN-BUILD MARKET

[C]   Other Markets

Page 20, add at end of subsection:

According to ENR's annual Top 400 Contractors report, design-build and EPC continue to grow as the project delivery method of choice in the international market. One example is U.S. contractor J.A. Jones, who reported 12 international contracting projects under way, 11 of which are design-build, turnkey projects, most in partnership with its design subsidiary, Lockwood Greene (*see* **§ 4.04**). *Revenue from Abroad Falls,* Engineering News Record, May 21, 2001, p. 109. On May 11, 2001, the Dutch government awarded a $2.3 billion design-build-finance-maintain (DBOM) contract for a new 100-kilometer high-speed rail project to Infraspeed, a Fluor Daniel-led joint venture. The civil work is being constructed under six separate design-build contracts. *Fluor-Led Group Wins Dutch Job,* Engineering News Record, May 21, 2001, p. 18.

Page 20, add new subsection at end of chapter:

[D]   Canada

Design-build is growing rapidly in Canada in the public and private sectors, a market that produced an estimated $65 billion in new construction in 2000. *What the World Spent on Construction,* Engineering News Record, Dec. 4, 2000, p. 38. A survey taken of its members by the Association of Consulting Engineers of Canada (ACEC) in 1999 found that while traditional contracts still hold the lion's share of revenues (84 percent), design-build accounted for 13 percent of total revenues, up from 12 percent in 1998. **Consulting Engineering in Canada,** Industry Profile, Association of Consulting Engineers of Canada (1998 & 1999). The same survey in 1998 showed that 42 percent of ACEC members had no design-build contracts, a figure that dropped to 33 percent in 1999, indicating a 9 percent increase among firms in just one year. Six percent of the ACEC members surveyed in 1999 said that design-build accounted for at least one-half of their revenue. For reasons not explained in the survey, firms located in Quebec were most likely to do design-build work, while Alberta companies were least likely to do it. The Canadian Design-Build Institute (CDBI) was formed in 1998 to help advance the best practices of design-build in that country. *See* **§ 2.08[B],** below.

Growth is reported in the Canadian government's use of design-build. Revenue Canada, Public Works and Government Services Canada, and the Department of National Defense are some of the federal agencies that are using design-build and bridging on their building projects. Use of design-build for the Revenue Canada Burnaby-Fraser Tax Services office in Surrey, British Columbia, cut the time in half for the new building that opened in March 1999. The award-winning project was the first major federal government building project in Canada using design-build and was completed ahead of schedule. Canadian Design-Build Institute Newsletter #2 (Aug. 1999). The Ministry of Natural Resources building in the city of Peterborough was a design-build project in which the Ministry developed an outline of its requirements, which included the total area of the structure and various structural loads. Four design-build contractors were prequalified and asked, as part of their proposal, to put together a team to complete the design of the structure and to provide a lump sum cost. The project did not go well, however, and the contractor sued its structural engineer for $700,000 for estimating errors. *See* **§ 23.01,** below. The U.S. General Services Administration (GSA) and Canada's Public Works and Government Services jointly requested technical proposals from design-build firms, due May 2001, for a $23–$26 million facility to be built in Alberta, Canada. The "shared border station" involves demolition of an existing station and construction of a joint facility for the United States and Canadian Border Service.

In Vancouver, British Columbia, 16 kilometers of the Millennium Skyway light rail extension have been built using the design-build method of delivery. This $140 million extension for the Rapid Transit Project 2000 (RTP) and SAR Transit is North America's largest precast segmental concrete guideway project. Since the entire project was erected in one year, it may also be the fastest such project. Work peaked at 1,800 employees working double shifts six days per week. *Vancouver Line Rolls On Despite Bumpy Start,* Engineering News Record, Feb. 19, 2001, p. 56.

THE HISTORY OF DESIGN-BUILD

Page 34, add new section at end of chapter:

§ 2.08 CANADIAN DESIGN-BUILD ORGANIZATIONS

[A] The Canadian Construction Association (CCA)

The Canadian Construction Association (CCA) has been way ahead of the design-build curve in terms of publishing guidelines on the process. In 1975, CCA published its ''Guidelines for the Design-Build Method,'' which included recommendations as well as CCA Document No. 14-1975, ''Agreement Between Client and Contractor with Contract Conditions of the Design-Build Stipulated Price Contract.'' This was several years ahead of such industry contracts in the United States. Then, in February 1986, CCA published a position paper titled ''Design Responsibility and the Trade Contractor,'' about the time that engineers and steel fabricators were addressing the same issue in the United States in the aftermath of the 1981 collapse of the Hyatt Skywalks in Kansas City. The paper noted that an increasing number of trade contractors are assuming responsibility for what had been considered in the past the exclusive domain of the design professional. The paper reported that ''[m]any trade contractors regularly contract for and successfully complete design-build projects'' and that it is a business decision whether a contractor wants to take on design responsibility. The paper observed that there had not been any shift in design responsibility and that the party who contracts to do the design is still responsible for it. As a result, CCA advised that contracts must be clear on who is doing what. CCA's paper noted that the increase in design work by trade contractors was directly related to increased complexity in design and that experienced installers may be in the best position to make design decisions. CCA's conclusion was that a guide on design responsibility is needed, directed not only to contractors but also to other parties in the design process.

CCA publishes design-build contract forms, discussed more fully in **§ 18.17**, below.

[B] Canadian Design-Build Institute (CDBI)

The Canadian Design-Build Institute (CDBI) is a special committee of the Canadian Construction Association (CCA), formed in 1998. The CDBI has five standing committees: Education, Standards of Practice, Owners, Membership, and Trade Contractors. In 2000, CDBI published a *Design-Build Practice Manual* that consists of two sections, 100 Series and 200 Series. Series 100, ''Introduction and General,'' covers topics such as Definitions, Overview of Design-Build, Design-Build Variants, Consultants, and Professional Regulation. Series 200, ''Procurement and Award,'' addresses developing RFQs and RFPs for design-build projects in sections entitled Content, Invitations to Respond or to Submit Proposals, Project Outline, Response Requirements, Evaluation, and Administration of the Selection Process. Membership information as well as the *CDBI Design-Build Practice Manual* can be obtained from the Institute at 400-75 Albert Street, Ottawa, Ontario K1P 5E7, phone (613) 236-9455, fax (613) 236-9526, or over their Web site at ⟨www.cdbi.org⟩.

CHAPTER 4

VARIATIONS IN DESIGN-BUILD

§ 4.08   THE MULTIDISCIPLINARY PRACTICE

Page 55, add at end of section:

According to an article in *Engineering News Record,* PricewaterhouseCoopers pays $50,000 or more to new hires in a construction-related field, with an annual bonus of $10,000. The company reportedly hires 4,000 to 6,000 new graduates annually, including 500 to 700 from design and construction programs. *College Recruiters Prowling for Talent,* Engineering News Record, Mar. 19, 2001, p. 44.

CHAPTER 5

DESIGNER AS SUBCONTRACTOR TO A CONTRACTOR

§ 5.08 CLAIMS BY OWNER DIRECTLY AGAINST DESIGN SUBCONTRACTORS

Page 70, add at end of section:

In some states an owner cannot make a claim directly against a subcontractor without a direct contractual relationship, whether the subcontractor is a trade contractor or a design professional. A direct claim is more likely when the owner is made a third-party beneficiary of the design-build subcontracts or when the design-builder assigns its rights against the subcontractor to the owner, as in the AT&T sample clause located in the **main volume.** When the prime design-build contractor does not carry professional liability insurance for design errors or omissions, an owner should insist that it has the contractual right to bring an action directly against the insured architect or engineer so that it can recover from the insurance proceeds. Owners seeking this right should be careful to coordinate the contract forms used, especially clauses dealing with dispute resolution. An owner who is required to arbitrate with the design-builder but litigate against the architect might find itself with inconsistent judgments and unable to fully recover all losses.

A Florida case decided in 2000 involved the issue of whether an owner could demand arbitration directly against the design-builder's architect. In *Cuningham Hamilton Quiter, P.A. v. B.L. of Miami, Inc.*, 776 So. 2d 940 (Fla. App. 3d Dist. 2000), the owner and contractor entered into a written design-build agreement for the design and construction of an entertainment complex. The design-build contract provided for arbitration of ''[a]ny controversy or claim arising out of or relating to this Agreement or its breach'' and stated that ''all parties necessary to resolve a claim shall be parties to the same arbitration proceeding. Appropriate provisions shall be included in other contracts relating to the Work to provide for the consolidation of arbitrations.'' The design-builder hired an architectural firm to provide architectural and engineering services for the project. The subcontract between the design-builder and architect also provided for arbitration by incorporation of the prime contract and AGC's Standard Form of Agreement between Contractor and Architect. When a dispute arose between the parties, the owner instituted arbitration proceedings against the design-builder but filed a lawsuit against the architect. Surprisingly, the architect sued to force the owner to arbitrate its disputes. The owner argued that it could not be forced to arbitrate its claims against the architect, since there was no signed contract directly between the parties. However, the Florida Court of Appeals ruled that a broad arbitration clause may include non-signatories and that since the owner's claims were ''intimately intertwined'' with the design-build contract, the architect was a necessary party to the arbitration and that even claims for negligent misrepresentation and fraud were related to the design-build contract and had to be arbitrated.

CHAPTER 6

BRIDGING

§ 6.01 WHAT IS BRIDGING?

Page 75, add at end of section:

The bridging concept is believed to have been conceived by George Heery, FAIA, formerly of Heery International, now a principal of The Brookwood Group in Atlanta, Georgia.

Page 75, add new subsection at end of page:

[D] Bridging in Canada

The concept of bridging has crept across the border into Canada. The Canadian Design-Build Institute (CDBI) publishes a *Practice Manual* that defines *bridging consultant* as: ''An individual or firm employed or engaged by an owner to develop a design to an advanced stage whereby the design-builder's role is reduced to completing the construction documents and construction (see Draw-Build).'' *Canadian Design-Build Institute Practice Manual,* 100 Series, Section 1.2.1 (2000). The *CDBI Practice Manual* defines *bridging* as:

> A form of design-build delivery whereby the owner enters sequentially into two separate contracts. The first contract is with a Bridging Consultant who prepares preliminary and developed designs. The second is with a draw-builder who completes (and assumes responsibility for the design prepared by the Bridging Consultant) the construction documents and builds and delivers the project.

Id. at Section 1.3. The CDBI plainly sees the role of the design-build contractor as merely drawing up the design prepared by the bridging consultant, calling this form of project delivery ''Draw-Build,'' while assuming full liability for the design prepared by the bridging consultant. The level of design done by a bridging consultant varies. In the best uses of bridging, there is a great amount of design discretion left up to the design-builder, who can be innovative in developing the design. When all that is left is to draw up the plans, however, the real value of design-build is lost.

CHAPTER 7

TEAMING FOR DESIGN-BUILD PROJECTS

§ 7.01 TEAMING RELATIONSHIPS

Page 83, add at end of section:

Some contractors feel that the key to successful design-build is to have a designer as an equity partner, taking on financial risk just like the contractor-partner. When the design firm's ability to make a profit is tied to the team's performance, there is more incentive to resolve design issues quickly and efficiently. Of course others are quick to point out that there is also added pressure that may compromise project quality. These conflicts of interest must be addressed by each design firm, balancing ethical considerations with potential for greater profits through an equity position in the project team. *See* **§ 24.01,** Conflicts of Interest, below.

§ 7.06 TEAMING AGREEMENTS

Page 89, add at end of section:

In September 2000, DBIA published a Design-Build Teaming Agreement Guide (Doc. No. 308), which gives guidance on the teaming concept, selection of team members, what an owner looks for in a team, organization of the design-build team, proposal preparation phase, post-award considerations, and termination of the teaming agreement. The *Guide* is well written and easy to read, and includes some sample language for various portions of the agreement.

CHAPTER 9

BUSINESS ISSUES FOR DESIGNERS IN DESIGN-BUILD

§ 9.09   INCENTIVE BONUSES IN DESIGN-BUILD CONTRACTS

Page 125, add at end of section:

On Salt Lake City's $1.6 billion expansion of I-15 project, all major lanes of traffic were opened five months earlier than the fast-track schedule's 4.5-year plan, and the job is reported still to be under budget. The design-build team anticipates receiving all of the $50 million at-risk bonus due to meeting all of the scheduled milestones. *Bulk of Ambitious $1.6-Billion I-15 Design-Build Job Complete*, Engineering News Record, May 14, 2001, p. 13.

CHAPTER 12

CONTRACTOR AND PROFESSIONAL LICENSING LAWS

§ 12.07 DESIGNERS BARRED FROM SUPPLYING MATERIALS TO JOBS THEY DESIGN

Page 179, add at end of section:

Like the Louisiana statute, discussed in the **main volume,** the Bylaws of the Architectural Institute of British Columbia (AIBC) in Canada state: ''An architect having a financial interest in any building material or device which the architect proposes to specify for a project shall disclose this interest to the client and shall request and receive written approval for such specification from the client and shall include a copy of this approval in the construction contract documents.'' Architectural Institute of British Columbia, Bylaw 32.7.

§ 12.09 INABILITY TO ENFORCE CONTRACT OR RECOVER FOR WORK

Page 181, add at end of section:

There are laws in Canada similar to those in Florida, Nevada, Arkansas, and Missouri. Under the licensing statutes of British Columbia, ''[a] person is not entitled to recover in a court a fee or remuneration for services rendered or work done contrary to this Act.'' Section 63.3, Architects Act of British Columbia, RSBC 1979, Chapter 19.

§ 12.13 POTENTIAL CRIMINAL LIABILITY

Page 185, add at end of section:

In an Arizona case, a contractor was convicted of a misdemeanor offense for acting as a residential contractor without a license. *State v. Wilkinson,* 10 P.3d 634 (Ariz. 2000). The accused was charged with acting as a contractor and advertising to provide contracting services without first obtaining a contractor's license. He was found guilty, assessed fines within the statutory range, and ordered to pay restitution to one owner of $22,429 and $22,365 to another. The court acknowledged that some owners choose to hire unlicensed contractors, but that even a contractor who reveals his unlicensed status to his client "would be no less guilty of the [crime] than one who affirmatively misrepresented the status of his license."

§ 12.15   COURT CASES DEALING WITH DESIGN-BUILD CONTRACTS

[B]   Cases Holding Design-Build Contracts Invalid

Page 189, add at end of subsection:

A 1998 Alabama case presented a host of licensing issues for the court dealing with an unlicensed design-build contractor and his scheme to get a building permit without a license. The court's opinion touches on nearly every topic in this chapter. In *Thomas Learning Center, Inc. v. McGuirk,* 766 So. 2d 161 (Ala. App. 1998), the owner of a day-care center entered into a written contract with McGuirk to design and build an addition to the day-care center. McGuirk told the owner that he was a licensed general contractor and that he had his own architect who would design the addition and who would be paid $390 from the $20,000 total fee for the project. The contract stated that a "drafting fee" of $390 had been deducted from the quoted $20,000 price to arrive at a contract price for the construction of $19,610. The job sat idle for weeks and the contractor told the owner that he was having problems obtaining a building permit from the city. Later, the owner learned that the contractor could not get the permit because he was not licensed. In Alabama, a general contractor's license is required for a commercial building project worth $20,000 or more. *See* Ala. Code §§ 34-8-1 *et seq.* [Note: The statute was amended in 1997, effective January 1, 1998, to increase the threshold to construction projects of $50,000 or more. The statute in effect before the amendment applies in this case.] Here, however, the contractor argued that no license was required since the contractor's cost after design fees would be just $19,610. However, while a state license might not have been required, a city business license was required in order to obtain a building permit. Lacking even the city license, McGuirk had a friend, Martin, who did have a city license, obtain the permit. The job fell further behind schedule and the owner went to Martin and demanded he take over and finish the "punchlist." At trial, the owner's expert witness testified that the plans and specifications for the project were deficient and contained numerous building code violations.

The case presented several licensing issues. The court noted that, under state law, a contract by an unlicensed general contractor is null and void as a violation of public policy, and that such contracts are illegal and unenforceable by the unlicensed general contractor. However, the owners did not take that position at trial. The court concluded that McGuirk's contract was actually for the full $20,000 rather than $19,610 and that he had violated state licensing laws by contracting without a contractor's license. The court noted that, "In a 'design/build' contract, the contractor is responsible for the project from design to construction and . . . assume[s] the responsibility for providing the necessary [designers]." As a result, the $390 drafting fee was included in the cost of the contract work for purposes of the licensing statute. The court concluded by stating that "[a] contrary holding would encourage unscrupulous contractors to avoid the requirements of the licensing statute by designating payments to subcontractors and suppliers as incident to separate contracts." Since a violation of Alabama's contractor licensing statute is a criminal offense, the court ruled that both Martin and McGuirk had broken the law. However, while their conduct might expose the contractor to criminal liability, the court held that such conduct did not in itself provide a civil cause of action for the owner. Nonetheless, the owner could sue for common law negligence to recover its damages.

In a 1998 case decided by Maine's Supreme Judicial Court, a design-build contract was held to be indivisible, for both design and construction, and totally unenforceable. The design-build contractor in *The Carvel Co. v. Spencer Press, Inc.,* 708 A.2d 1033 (Me. 1998) submitted a cost proposal for an office expansion project but the owner told the contractor that it would put the work out for bids. Nonetheless, the contractor later delivered a set of unsealed plans and specifications to the owner with a cover letter stating that its proposal was solely as a design-build contractor, not as a consulting engineer that produces plans, and stating that the design was for the contractor's internal use only and not to be distributed to its competitors for bidding. The owner later put the job out for bids as a design-build project and found the original contractor's price to be high. The contractor charged $36,896 for the value of the design work but the owner refused to pay. The trial court awarded the design-builder $25,000 plus attorneys' fees and interest, but that decision was reversed on appeal. The appellate court held that the contract was not divisible and, since the plans and specifications were not sealed, they were of no value to the owner. The contractor was barred from recovering—not because it was unlicensed, but because the design documents were not sealed.

Page 194, add new section at end of chapter:

§ 12.19 CANADIAN LICENSING LAW REQUIREMENTS

In Canada, as in the United States, a firm, corporation, or partnership may practice professional engineering in the name of the firm if it is the holder of a "permit" or certificate issued to it under the licensing laws. With the exception of Ontario, no other Canadian province has defined "design-build" in the architectural licensing laws or codes of ethics. The Ontario Association of Architects defines *design-builder* as "a person who is in the business of construction, enlarging, or altering buildings and who engages or retains a holder [i.e., of a certificate of practice to practice architecture] to provide architectural services in connection with a project for the construction, enlarging or altering of a building." Regulation

27, Architects Act, R.S.O. 1990, Ch. A26. The following subsections show some of the similarities and differences between United States and Canadian licensing law as relates to practice by corporations and partnerships.

[A] Corporations

The requirements for obtaining a corporate permit in Canada are similar to those of many American licensing boards, i.e., that "at least one member or licensee who is a full-time employee, partner or officer of the firm" must be appointed to assume responsibility for the professional conduct of the firm. *See, e.g.,* Engineering, Geological and Geophysical Professions Act of the Northwest Territories, R.S.N.W.T. 1988, as amended; Permits, Section 20. *See also* Consolidated Statutes of Ontario, c. A.26, s.12. The percentage of control required varies among the provinces, much as it does among the states, with ownership requirements ranging from 50 to 100 percent of the owners or partners who must be licensed architects or engineers.

For example, in Alberta, an "architectural firm" is one in which registered architects hold a majority interest and control the corporation. Section 17, Alberta Architects Act, Chapter A-44.1, Revised Statutes of Alberta 1980. The requirement for engineers is identical, i.e., engineers must hold a majority interest and control the corporation. Likewise, in Manitoba and Nova Scotia the majority of the voting shares must be held by practicing architects, licensed in that province, and the majority of the directors must likewise be licensed architects. Section 16, Manitoba Architects Act, Chapter A130, R.S.M. 1987; Section 21, Nova Scotia Architects Act 1968. However, on Prince Edward Island, it is required only that one of the directors be an architect whose ownership interest is no less than any other shareholder or director. Section 14(4), Architects Act of Prince Edward Island, Chapter A-18.1. If, however, the corporation is a non-resident, then at least two-thirds of the directors must be licensed architects and a majority of the voting shares must be owned by architects. Section 14(5), Architects Act of Prince Edward Island, Chapter A-18.1. Under the Ontario Architects Act, in order to obtain a corporate certificate to practice, the majority of the directors of the corporation must be composed entirely of architects or engineers licensed in Ontario, and a majority of each class of stock in the corporation must be owned by such persons. In addition, all other shareholders must be full-time employees of the corporation, the primary function of the corporation must be to engage in the practice of architecture, and at least one director or full-time employee must be personally responsible to supervise and direct the practice of architecture. Consolidated Statutes of Ontario, c. A.26, s.12. Non-licensed persons cannot own, control, or exercise direction over (individually or together with others) more than 10 percent of the total number of shares issued and outstanding. *Id.,* c. A.26, s. 21. As a result, if a design-build team in Ontario is made up of a contractor and an architect who form a new entity, the architect must own at least 90 percent of the design-build firm in order for the corporation to legally practice architecture.

Other provinces, however, are less restrictive, and in Newfoundland a corporation may be issued a Certificate of Approval to practice architecture if only two-thirds of the directors of the corporation are licensed design professionals and not less than 51 percent of the shares of stock are owned by those directors, or 100 percent of the shares are owned by architects licensed in Newfoundland. Sections 23, 24 of the Newfoundland Architects Act, Chapter 64. Likewise, in British Columbia, a corporation can be registered to practice architecture if a simple majority of each class of voting shares of the corporation is owned by architects, the majority of the corporate directors are architects, and the chief executive officer is an architect. Section 24.1, Architects Act of British Columbia, RSBC 1979, Chapter 19. This still slants ownership to the architect partner in a design-build team that forms a new corporation to hold the prime design-build contract and lawfully practice architecture.

[B] Partnerships

As with corporations, the Canadian licensing laws restrict partnerships that may practice architecture with several variations in ownership requirements. For example, in Newfoundland, two-thirds of the partners must be licensed design professionals (architects, engineers, planners, etc.). Sections 23, 24 of the Newfoundland Architects Act, Chapter 64. The law of Ontario is more restrictive and a partnership may obtain a certificate of practice only if all of the partners are members of the Ontario Association of Architects, i.e., licensed architects. A partnership may also be comprised of corporations if one or more of the corporations holds a certificate to practice architecture and the other corporations hold general certificates of authority issued under the Professional Engineers Act. Consolidated Statutes of Ontario, c. A.26, s.16. The licensing laws of British Columbia and Nova Scotia are similar to those of Ontario and, other than partnerships with engineers, the licensing law strictly prohibits partnerships with non-architects. Section 63, Architects Act of British Columbia, RSBC 1979, Chapter 19; Section 18, Nova Scotia Architects Act 1968. As a result, in these provinces a joint venture partnership between a contractor and an architectural firm would not be permitted. There is no statutory authority to issue a certificate of authority to a partnership between an architect and a contractor. If Canadian courts should ever interpret design-build as constituting the practice of architecture, architects and engineers will be the only ones who can qualify for a certificate of authority in most cases, and even a partnership between an architectural firm and a contractor would not meet the qualifications to practice architecture.

[C] General Requirements

As in the United States, Canadian licensing laws have not yet caught up with the growth of design-build. The Canadian Design-Build Institute (CDBI) recommends that anyone considering the design-build procurement method in

Canada consult with the appropriate province's architectural and engineering associations. According to *CDBI Practice Manual,* 100 Series, Section 5.2 (2000), the provincial associations generally agree on certain base issues regarding the use of design-build:

1. *Avoid Conflicts of Interest.* A consultant must avoid actions and situations where there is, or appears to be, conflict between the consultant's personal interests and professional obligations to the public, the client, and other consultants.

2. *Obtain Written Consent.* A consultant can provide professional services to more than one client on the same project by written agreement by all interested parties prior to providing services to the second and subsequent parties. (CDBI notes that, ''This allows the Bridging Consultant or Criteria Consultant to the owner to become the design-builder's consultant.'')

3. *Remain Impartial.* A consultant, acting as interpreter of the construction contract documents and reviewing construction for conformance with the contract documents, must render decisions impartially when performing professional services such as field review, certifying payment, and certifying progress of the work. CDBI states that, ''Services must be rendered as fully as when engaged by a third-party client. Financial interests must not override professional responsibilities.''

    There is some disagreement on the point between the CCA and the Association of Consulting Engineers of Canada (ACEC). The engineers object to CCA's Design-Build Document 14 because of a perceived conflict of interest in advising the owner. In an October 2000, letter to its members, ACEC warned that inexperienced owners will rely on the consultant to provide advice during the project and that while the consultant in a design-build setting works for the design-builder, there may be a duty of care to the owner owed by the consultant. ACEC recommends making clear in the contract the limitations on the consultant's role. What should not change, however, in either setting is the design professional's duty to remain impartial. Whether working for an owner or for a contractor, design professionals may be called on to make a decision that is against the financial interests of their client. Architects and engineers do this routinely while being paid by the owner and there is no reason that they cannot remain impartial when being paid by the contractor. Design professionals do not ''check their ethics at the door'' when entering into design-build contracts.

[D]   Design-Led Design-Build

In Canada, a design-builder may generally be a design professional as long as the professional also provides the architectural-engineering design or the construction contract documentation services on the project. CDBI states that it is generally agreed among the licensing authorities that the design professional's "dual role" must be disclosed to (and acknowledged by) all the project's contracting parties and authorities having jurisdiction, and that the professional provides impartial professional services. CDBI recommends that "[d]isclosure should be made at the earliest opportunity, and recorded in the consultant's construction documents and permit application forms."

The Architectural Institute of British Columbia addresses designer-led design-build in its Bylaw 31.5, which states: "An architect may be a project's contractor, of the architect's own design and/or construction contract documents." The Bylaw goes on to state, however, that architects acting as the project's contractor must fully disclose in writing such status to all of the project authorities and receive their written acknowledgment. Official commentary to that Bylaw explains that the architect-contractor role is limited to projects designed by the architect, and that in this role the architect "must render architectural services as fully and impartially and must be as disinterested as an architect who is solely serving a third-party client."

Disclosure is specifically required in Nova Scotia's Code, which states: "An architect shall not be financially interested in the bids as or of a contractor on competitive work for which he is employed as an architect unless he has the consent of his client or employer." Alberta regulations state that an architect "who engages in another profession, business or occupation concurrently with the practice of architecture must not allow that outside interest to jeopardize his professional integrity, independence or competence." Rule 6, Outside Interests and the Practice of Architecture, Code of Ethics, Alberta Regulations 240/81. So for architects entering into design-build in Canada, while it may seem pointless to state the obvious, provincial ethical rules require that the firm advise the owner up front that the architect is acting as builder and seek the client's consent to serve in that dual role.

CHAPTER 13

INSURANCE AND BONDING FOR DESIGN-BUILD

§ 13.21   VALUABLE PAPERS INSURANCE

Page 212, add at end of section:

This type of policy not only covers the cost to reconstruct the papers and records but also the cost to research the last information. Such policies cover papers and records that belong to the insured and those in the insured's care, custody, and control. Data processing media are excluded as are "electrical or magnetic injury, disturbance or erasure of electronic recordings." To fill the gap, some insurers recommend an Electronic Data Processing (EDP) policy that covers equipment and electronic media such as disks or tapes on which data is stored. It is possible to endorse the EDP policy to cover valuable papers as well, giving the insured the convenience of having a single policy and single insurer for both coverages.

CHAPTER 14

PUBLIC SECTOR DESIGN-BUILD

§ 14.06   COMPETITIVE BIDDING LAWS

[B]   Cases Finding No Violation of Competitive Bidding Laws

Page 229, add at end of subsection:

Oregon passed specific legislation that exempts public contracting agencies from competitive bidding if local contract review boards make findings that certain conditions have been met. ORS 279.015(1); *see* Oregon Public Procurement Laws, **Appendix E** in the **main volume.** The public agency may use ''alternate contracting and purchasing practices'' by statute in those cases, which would include use of design-build. That statute was put to the test in *Associated Builders & Contractors, Inc. v. Tri-County Metropolitan Transportation District,* 12 P.3d 62 (Ore. App. 2000). In that case, the Tri-County Metropolitan Transportation District in Portland, Oregon, issued a decision exempting from competitive bidding the contract to construct a light-rail extension to Portland International Airport. Associated Builders & Contractors, Inc. challenged that decision and lost. The dispute arose when the District entered into negotiations with Bechtel Infrastructure Corporation for a $125 million design-build contract without competitive bids. The Court of Appeals noted that, ordinarily, public contracts must be awarded on the basis of competitive bids, which would mean that the contract to construct the light-rail extension would be awarded in that manner. The court found, however, that the District had complied with the statute permitting a public agency to exempt a contract from competitive bidding by employing an ''alternative contracting method.''

[C]   Design-Build Held to Violate Competitive Bidding Laws

Page 229, add at end of subsection:

A 1999 case from the Virgin Islands held that the process used to hire a design-build contractor for a $25 million jail project violated competitive bidding laws and, as a result, an injunction was granted to prevent the award of the contract. In *C&C/Manhattan v. Government of the Virgin Islands,* 1999 V.I. LEXIS 4, the government hired Heery International to prepare a 70-page RFP for design-build proposals for a new jail on St. Croix, with a weighted selection scale to evaluate the proposals. After reviewing three proposals, Heery recommended award to the team of Hyde Park/Perini for a negotiated contract. C&C, the low bidder, sued to stop the award, arguing that by law the contract must be awarded to the low bidder. The court rejected the government's arguments that the contract was a professional services contract that could be negotiated, stating that after the new jail was designed, the contractor would be primarily responsible for actually constructing correctional facilities that accounted for about 70 percent of the contract work. The court regretted its ruling, stating that the government was merely trying to find a way to expedite the design and construction of a much-needed facility. However, the court ruled that although design-build is authorized for the federal government, "it is for our Legislature, and not the courts, to decide whether and under what circumstances it should be authorized in the Virgin Islands."

§ 14.09   LEGISLATIVE DEVELOPMENTS

Page 232, add at end of section:

In the 2001 state legislative session, there were 50 bills introduced in 19 states across the United States dealing with private and public use of design-build. By far, the majority of the bills were aimed at permitting states and state agencies to use design-build either across the board or on limited "pilot" projects for schools, highways, airports, jails, and other state projects. Minnesota led the nation with nine design-build bills, followed by California with seven, Texas with six, and Missouri with five. As of May 15, 2001, only five of these bills had been *passed*:

> **Kentucky: H.B. 347** (permits architects to design same project they build);
> **Virginia: S.B. 1049** (permits the state Transportation Board to award five design-build contracts at up to $20 million per year and five over $20 million);
> **Washington: H.B. 1680** (permits the state department of transportation to use design/build for projects over $10 million) and **S.B. 5060** (amends current design-build laws, expands types of public bodies authorized to use design-build);
> **West Virginia: S.B. 298** (to continue authorization of the legislative rule enacted March 23, 2000, which permitted the department of administration to select design-build contractors, 148 CSR 11).

These bills are included in **Appendix E, this supplement.** Others may have been enacted after May 1, 2001, and those will be reported in future supplements.

The topics covered by the bills were very similar: 63 percent deal with state projects (half of which are related to state highway and transportation projects) and 12.5 percent deal with cities and counties. Four bills authorize studies on design-build, three focus on school districts, and two others deal with state licensing laws for design-build contractors.

§ 14.10 PREPARING AND RESPONDING TO RFQs AND RFPs

Page 233, add at end of section:

The *Practice Manual* developed by the Canadian Design-Build Institute (CDBI) also has guidelines for developing Requests for Qualifications for design-build projects. Included in CDBI's recommendations are Aims and Objectives of an RFQ/RFP, Content of a Typical RFQ/RFP, Invitation to Respond to an RFQ (or to submit proposals), Project Outline, Response Requirements, Evaluation, and Administration. *Canadian Design-Build Institute Practice Manual,* 200 Series, Section 2.0, Developing RFQs (2000).

§ 14.15 PAYMENT OF STIPENDS

Page 243, add new subsection after [F]:

[G] Payment of Stipends in Canada

Two "Super Jail" projects for the Ontario Realty Corporation generated debate over the reasonableness of stipends paid to the proposers. This information was reported in the Canadian Design-Build Institute's Practice Bulletin #2 (Feb. 1999). The first RFP provided an honorarium of $35,000 although design-build teams reported costs ranging from $200,000 to $500,000 to respond, with architectural consultants incurring $100,000 to $200,000 alone. The second project had similar conditions but, after one of the pre-qualified participants withdrew, the owner agreed to divide that team's honorarium among the remaining teams. In the aftermath, the Ontario Association of Architects and the Ontario General Contractors Association met with the owner to voice their opinions that the first two projects were out of proportion to the honoraria paid. The two groups urged that in future projects the honoraria be set at one percent of the budgeted cost of the project or, alternatively, that all teams carry a "cash allowance" in their price that would reflect the amount that would be paid to the unsuccessful teams (i.e., out of the contract award price), or that the requirements for the submissions be significantly reduced. As a result of this input, the next RFP had increased honoraria and reduced submission requirements.

The Canadian Design-Build Institute (CDBI) stated in its February 1999 *Practice Bulletin* that the subject of honoraria for design-build RFPs ("proposal calls") has been one of the most controversial issues in design-build, noting candidly: "Generally the honoraria is [sic] no more than a token payment, that at best may cover reprographic costs, but little else." CDBI observed, however, that even with no honorarium, some proposers will elect to participate without compensation as a cost of doing business. The topic of stipends or honoraria is addressed in the *CDBI Practice Manual* published in 2000. The *Practice Manual* defines *honorarium* slightly differently from *stipend*. The *Manual* defines *honorarium* as: "The amount paid to proponents to *offset their costs* in responding to an RFP." *Canadian Design-Build Institute Practice Manual,* 100 Series, Section 1.2.1 (2000). The *Manual* defines *stipend* as: "The amount paid to proponents to *contribute to the costs* of responding to an RFP." The *Manual*'s Section 8.0, Payment for Proposals, alerts owners to the fact that design-build proposals are the result of considerable design work performed over many long hours and that "the owner derives added benefit from this effort and should therefore be prepared to pay for these services." *Canadian Design-Build Institute Practice Manual,* 200 Series, Section 8.1 (2000). CDBI states that the level of detail required by the proposal call and the complexity of the project should impact the compensation paid to proposers.

A Canadian joint industry and federal government design-build task force studied the issue of stipends and in September 2000 issued the following recommendation: "The remuneration to each proponent for the preparation and submission of a design-build proposal should be at least 50% of a reasonable estimate of the costs for the submission, including professional fees and the design-builder's costs and expenses." Whether public or private owners will accept the recommendation remains to be seen. As in the United States, stipend amounts vary and Canada has no law that sets the amount of compensation to be paid as a stipend on design-build projects.

CHAPTER 15

DESIGN-BUILD AND THE FEDERAL GOVERNMENT

§ 15.03 THE DEPARTMENT OF DEFENSE

Page 251, add at end of section:

The U.S. Air Force awarded a design-build contract to Harkins Builders for 60 two- and three-story town homes at Bolling Air Force Base in the District of Columbia. Another military housing project is being done by the U.S. Naval Facilities Engineering Command, which solicited proposals in April 2001, for a $17.9 million design-build project to replace family housing at Naval Air Station in Lemoore. The Navy's solicitation described the project as including the design and construction of 100 new family housing units and the demolition of 100 existing housing units, plus pavement and utilities.

§ 15.04 THE CORPS OF ENGINEERS

Page 251, add at end of section:

In October 2000, Congress clarified federal design-build rules in a report dealing with the U.S. Army Corps of Engineers' use of design-build on five civil works pilot projects, stating that the Corps is to use the two-phase design-build selection process enacted by the Federal Acquisition Reform Act (FARA) of 1996. S.2796, Sec. 221, Design-Build Contracting, 146 Cong. Rec. H11624 at H11631 (Oct. 31, 2000). The Corps of Engineers solicited proposals in November 2000 for a new $30 million athletic facility at the U.S. Air Force Academy in Colorado. The design-build project will have offices, meeting rooms, football locker room, weight training area, and educational space. The Corps is also using design-build to renovate a portion of Hanscom Air Force Base in Massachusetts for 95,000 square feet of administrative office space. In Dayton, Ohio, the Corps solicited proposals in December 2000 for a new hangar and taxiway at Wright-Patterson Air Force Base. The design-build project includes a 235,000 square foot hangar to house about 60 planes, a tower to showcase ballistic missiles, and an 80-foot-wide connector building. In May 2001, the Corps solicited proposals for design-build of a new Defense Equal Opportunity Management Institute facility, a two-story building estimated at over $10 million, at Patrick Air Force Base.

§ 15.06   FEDERAL BUREAU OF PRISONS (FBOP)

Page 252, add at end of section:

The Bureau awarded a $129 million design-build contract to P.J. Dick, Inc. for a 960-bed, high security prison in Preston County, West Virginia, on a 1,036-acre site just off Interstate 68. The project includes an adjacent minimum security camp for up to 300 men, with completion slated for 2003.

§ 15.12   PROHIBITION AGAINST DESIGNER-LED DESIGN-BUILD

Page 254, add at end of section:

This type of law barred an engineering firm from competing for a design-build contract for the city of Phoenix's Lake Pleasant sewage treatment plant in late 2000, because the firm had done technical studies for the city prior to the city's decision to use the design-build delivery system on the project. *Phoenix Court Hears Bid Protest,* Design-Build Dateline, Nov. 2000, p. 11. The city took the position that under Arizona design-build law, the city's contract selection committee could not include on the short list any person or firm that included or employed any person or firm that had provided for compensation any services relating to the project. *See* A.R.S. § 34-603.C.2(f), **Appendix E, main volume.** DBIA testified in support of the engineering firm, arguing that when preliminary work is performed by a consultant, and the delivery method is changed, the consultant should not be excluded from the work, provided that the information provided by the consultant is made available to all bidders so there is no unfair competitive advantage.

In 2001, Kentucky amended its state procurement laws to prohibit local governments from hiring one firm to provide both architectural services and construction management services on the same project. However, the amendment expressly exempts design-build projects where, presumably, an architect could provide the design and then manage the trade contractors as a construction manager. *See* Ky. H.B. 347 (2001), **Appendix E, this supplement.**

CHAPTER 16

OWNERSHIP OF DESIGN PROPOSALS, TECHNICAL SUBMISSIONS, AND COPYRIGHTS

§ 16.08 COPYRIGHT INFRINGEMENT AND REMEDIES

[G] Sample Cases

Page 268, add at end of subsection:

Be careful when an owner, developer, or design-build contractor asks an architect or engineer to finish plans prepared by another design professional. Taking over another designer's project usually means trouble in the prior relationship and there may be an angry design firm wanting to protect its copyrights if the design-builder gives the design to someone else to use. Those were the facts in a 2001 Maryland case, *Nelson-Salabes, Inc. v. Morningside Holdings of Satyr Hill, L.L.C.,* 2001 WL 419002 (D. Md. 2001). In that case, a developer bought a piece of land for use as an assisted living facility. Of particular interest to the developer was the fact that the seller had architectural plans prepared by an architect that were already approved by the county. The developer then hired a design-build contractor, who hired a second architect to further develop the approved design. The original architect sued the developer for copyright infringement based on its registered copyright in the footprint and front exterior elevation. The design-build contractor and its architect were not sued. The original architect was awarded $736,037, representing the developer's profit from use of the infringed plans. This suit could just as easily have included the design-build contractor and the second architect who used the copyrighted plans without permission of the copyright owner.

In yet another design-build case, where an architect's design was used by others, the architect sued multiple defendants for copyright infringement, breach of contract, conversion of personal property and work product, and for the reasonable value of the drawings. In *Bush v. Laggo Properties, L.L.C.,* 2000 WL 1763409 (Ala. App. 2000), the architect had been hired by a design-build contractor as part of a team for certain buildings. After performing some design work, the design-builder told the architect that negotiations on the project had ceased because the project was over budget. Later, the architect learned that a different contractor had obtained a building permit and was using the same architectural plans to construct a building. The plaintiff-architect sued for breach of contract, claiming that he was an intended third-party beneficiary of the prime design-build agreement between the design-builder and others. The court agreed that the architect could sue as a third-party beneficiary for breach of contract.

Page 272, add new section at end of chapter:

§ 16.13   COPYRIGHT ISSUES IN CANADA

The issue of patents and copyrights is dealt with in Canadian design-build contracts in a manner similar to the AIA, AGC, and DBIA approaches. The CCA/CSC/RAIC design-build Document 14 (2000), states in ¶ 1.1.11: "Copyrights for the design and drawings prepared by or on behalf of the Design-Builder belongs to the Consultant or other consultants who prepared them." *See also* ¶ 4.2 of Document 15 (2000). The *CDBI Practice Manual* similarly states:

> Patents and copyrights belong to their creators. Owners should be aware that design patent rights remain with the patent holder and copyrights normally remain with either the designer or the design-builder, or both. Payment of a proposal fee does not confer a right on an owner to use copyrighted or patented material.

Canadian Design-Build Institute Practice Manual, 200 Series, Section 9.1 (2000). As in the United States, a design firm can give up its copyrights by written agreement. Normally, this is done in exchange for some protection by the other party for claims arising out of subsequent use of the design.

When the design-builder grants the owner rights to use the copyrighted design, CDBI recommends that: (1) remuneration should be reasonable and sufficient; (2) copyright when transferred should be honestly replicated; (3) liability for errors and omissions inherent in the design of copyrighted material should be addressed and/or transferred; (4) the author should receive acknowledgment and credit for the design; and (5) professional ethical standards should be maintained.

CHAPTER 17

DESIGN-BUILD PROJECT TYPES AND PROJECT STUDIES

§ 17.02 TRANSPORTATION PROJECTS

[C] Highways

Page 280, add at end of subsection:

The Massachusetts Highway Department awarded a contract to a contractor-led team of Modern Continental Construction Co. and URS Corp. for a $385 million design, build, finance, operate, and maintain project, consisting of a 21-mile stretch of Route 3 from I-95 in Burlington to the New Hampshire border. URS is providing the engineering design. The project involves adding one lane in each direction and a full shoulder adjacent to the high-speed lanes, with replacement of bridges and other improvements.

The biggest design-build project in the nation will be the $1.67 billion Southeast Corridor project in Denver, Colorado, nicknamed "T-Rex." The Colorado Department of Transportation originally had three prequalified teams, but Valley Highway Constructors, led by Bechtel Infrastructure Corp., dropped out of the running in October 2000, leaving Southeast Corridor Constructors, led by Peter Kiewit & Sons, Inc., and Valley Corridor Constructors, led by Flatiron Structures Co. On June 1, 2001, the contract was awarded to Southeast Corridor Constructors. Kiewit was the lead contractor on Salt Lake City's $1.59 billion I-15 project, previously the largest design-build project in the nation. The Denver project includes 17 miles of improvements to Interstates 25 and 225, plus 19 miles of new double-track light rail, with 13 new stations, serving downtown Denver. The project was designed by bridging to a 30 percent level prior to contract award to the design-build team. The paperless RFP was issued in November 2000, on a secure Web site at ⟨www.southeast-corridor.com⟩ and on CD-ROM. Completion is scheduled for fall of 2006.

Also in Denver, a $233 million design-build contract was awarded to MKK Constructors, a joint venture between Washington Group and Kiewit Western Co., as part of Segment IV of E-470, the new tollway around the east side of the city. Other projects include the new City Hall, an aquarium, New Mile High stadium (Invesco Field), Pepsi Center arena, and a variety of private sector projects, jails, and hospitals in and around the Denver area.

Maryland's Department of Transportation awarded a contract for a $9.5 million reconstruction of the Baltimore Beltway (I-695) as a design-build project led by T.C. Simmons, Inc. The project is just 1.25 miles, but involves construction of a Jersey barrier wall, access ramps, and noise barrier, and overlaying the existing section of the Beltway with asphalt.

In Florida, a $5.8 million design-build welcome center contract was awarded by the Florida Department of Transportation to G.A.C. Contractors. The 13,000 square foot facility is located on Highway 231 near Campbellton and has its own wastewater treatment plant. Construction is planned for completion in spring of 2002.

§ 17.04   WATER TREATMENT PLANTS

Page 283, add at end of section:

A design-build team of Black & Veatch and Dillingham Construction was awarded a $30 million contract to build a 20 mgd water treatment plant for the California Water Service Co. of San Jose. The project is slated for completion in 2003.

§ 17.06   GOVERNMENT BUILDINGS

Page 284, add at end of section:

In April 2001, the State of California, Department of General Services, put out a request for proposals to design-build contractors for a $24.9 million, six-story, 168,000 square foot state office building to be built in San Diego.

The City of Aurora, Colorado, selected the design-build team of The Weitz Co. and Michael Barber Architecture to design and build its new $66.5 million Municipal Center, to be located next to the Aurora Municipal Justice Center. The contract was awarded after a year-long process of evaluating proposals and bids. The project is planned for completion by the end of 2002.

The National Aeronautics and Space Administration (NASA) solicited proposals in December 2000 for a $20 million design-build project to demolish two existing buildings at the George C. Marshall Space Flight Center in Huntsville, Alabama, and to construct a new office building on the site.

Another federal agency, the United States Postal Service, has increased its use of design-build. Recent projects include a $6 million, 54,000 square foot addition to the Postal Service Headquarters in Arlington, Virginia, awarded to Interwest Construction of Salt Lake City, Utah, and an 88,000 square foot addition to the Postal Service's processing and distribution center in Champaign, Illinois, awarded to The Korte Company of St. Louis, Missouri. The Korte Company was also the contractor on a $21 million design-build renovation project for the U.S. Postal Service in Daly City, California, where the design-build team renovated a 52-year-old, 250,000 square foot food distribution warehouse into an International Service Center for the Postal Service. The scope of work included asbestos re-

moval, lead-based paint abatement, and structural upgrades to meet seismic standards. Use of design-build reportedly cut project delivery time from 14 months
to 7 months. *Design-Build is Growing,* The Military Engineer, Jan.-Feb. 2001,
p. 19. The Postal Service has also awarded a $2.3 million design-build contract
to C.D. Barnes Associates, Inc. for a new 28,780 square foot carrier annex in
Kentwood, Michigan.

§ 17.07   ENTERTAINMENT AND SPORTS

Page 285, add at end of section:

The Arizona Cardinals' football stadium should be ready for the 2004 NFL
season if the project's design-build contractor, Hunt Construction Group, stays
on schedule. The $335 million project in Tempe will have a 240 foot wide retract-
able roof. The project was defeated in 1998 by voters, but passed by a narrow
3.7 percent margin in November 2000.

In Pittsburgh, Pennsylvania, the Sports & Exhibition Authority has under-
taken a billion-dollar construction program that includes two major league sports
stadiums (for the football Steelers and the baseball Pirates), a new convention
center, and associated infrastructure projects. Three Rivers Stadium was razed in
the process. Barton Malow/Dick Corp. is the joint venture acting as the design
and construction manager for the project, named PNC Park. Given the time con-
straints of the owners, design-bid-build was out of the question and the project
had to be fast-tracked. Design-build was selected by the Pirates as the only way
to meet the project's goals, and bridging was selected to give the owner more
control. Using a process referred to as the ''Prose Process,'' when the GMP documents were issued, the conceptual architect (the bridging consultant) prepared a
detailed Prose Statement indicating on a drawing-by-drawing basis those elements
that remained incomplete. The Prose Statement then outlined the quality and quantity that would ultimately appear in the construction documents when final. The
design-build contractor's statement of GMP Qualifications was then based on the
Prose Statement, discussed in a line-by-line-item basis. The result of that discus-
sion became part of the GMP contract. The process reportedly increased all parties' confidence in the accuracy of the GMP documents and the GMP pricing.
Big Building Boom in the 'Burgh, Under Construction, ABA Forum on the Construction Industry (March 2001).

The $35 million Fonner Park Exposition and Events Center in Grand Island,
Nebraska, is a design-build project awarded to Kiewit Construction Co. as contractor and Leo A. Daly as designer and architect of record, in association with James
Cannon & Associates. The facility will have a 113,200 square foot multipurpose
arena seating 7,500 for sporting events, conventions, concerts, and trade shows,
plus renovation of the existing grandstand structure.

A $20 million design-build contract has been awarded to C.D. Barnes Asso-
ciates, Inc., for the new Celebration Village Cinema in Grand Rapids, Michigan.
The complex will consist of a new 18-screen state-of-the-art cinema, including

an IMAX theater, a retail shopping center, and office space. The 159,000 square foot project is part of a larger $155 million development project.

The Georgia Institute of Technology has awarded a $37 million design-build contract to Beers Construction for a student athletic facility, designed by Hastings and Chivetta Architects of St. Louis, Missouri as a subcontractor to Beers.

§17.08 MUSEUMS AND OTHER ATTRACTIONS

Page 286, add at end of section:

In Seattle, a technically difficult design by Rem Koolhaas for the city's new $159 million Central Library caused the city to let a separate design-build contract for the glass curtain wall of the 15-story building. Requests for proposals were sent to six short-listed contractors in November 2000. Hoffman Construction Co. is the project's general contractor/construction manager. *City Leaders Take Design-Build Approach,* Engineering News Record, Nov. 20, 2000, p. 9.

A new casino for the Tuolumne Band of Me-Wuk Indians near Sonora, California, was awarded to Kitchell Contractors and KGA Architecture as the design-build team. The project includes 20,000 square feet of pre-engineered building in phase 1, with phase 2 including a two-story 126,000 square foot structure for casino, restaurant, gift shop, offices, and other facilities.

§ 17.10 EDUCATIONAL FACILITIES

Page 288, add at end of section:

The University of California Irvine solicited bids from prequalified design-build teams in March 2001 for a three-story, 53,000 square foot, $17 million research laboratory facility for the School of the Physical Sciences. Another laboratory using design-build is Phase 2 of Nano Science Laboratory at the Naval Research Laboratory Center, a $10 to $15 million design-build project being built in the District of Columbia.

Projects at two elementary schools in Florida were awarded to The Morganti Group, Inc. and BRPH Architects-Engineers as the design-build team. The West Gate Elementary School modernization and replacement is a $10.3 million project in West Palm Beach, while the Miramar E6 elementary school in Miramar is an $11 million design-build project. Both new schools are 124,000 square feet each.

§ 17.11 MANUFACTURING PLANTS

Page 289, add at end of section:

Anheuser-Busch Co. awarded a $6.8 million design-build contract to HBE Financial Facilities to update and expand their employees' credit union headquarters facility in St. Louis, Missouri. The project has a 17-month schedule for a 32,000 square foot addition to the existing headquarters building.

Page 291, add new section at end of chapter:

§ 17.14   HOSPITALS

Phipps/McCarthy was the joint venture design-build contractor for the $110 million University Hospital Anschutz Outpatient Pavilion at University of Colorado Hospital. The 450,000 square foot, seven-story building was the largest design-build health care project in the country, according to architect Rob Davidson of H&L.

Another hospital project is the $35.1 million expansion and renovation of Sun Health Del E. Webb Memorial Hospital, which is being done via design-build with McCarthy as lead contractor. The project includes 220,000 square feet of new construction, including a six-level tower and other additions to the hospital.

DESIGN-BUILD CONTRACTS

Page 308, add new section at end of chapter:

§ 18.17   CANADIAN CONTRACT FORMS

[A]   Canadian Construction Documents Committee (CCDC) Forms

There is a joint drafting committee made up of members from the Association of Consulting Engineers of Canada, the Canadian Construction Association, Construction Specification Canada, and the Royal Architectural Institute of Canada who, like their American counterparts in EJCDC, *see* **§ 18.04, main volume,** have formed a joint committee to prepare industry forms. The group calls itself the Canadian Construction Document Committee (CCDC). In 2000, the CCDC published Document 14, Design-Build Stipulated Price Contract, and Document 15, Design-Builder/Consultant Contract, but without the endorsement of the Consulting Engineers, as explained in **§ 18.17[B], this supplement.** Here are some of the highlights of these forms:

Currency. For obvious reasons, ¶ 4.4 of Document 14 makes it clear that all amounts stated in the Contract Price are in Canadian funds.

Interest. Interest on late payments is compounded monthly and is to be set by the parties at an agreed percent above the prime rate of interest quoted by the Royal Bank of Canada. ¶ 5.3 of Document 14; ¶ 4.3 of Document 15.

Language. Both agreements force the parties to select which language controls, either French or English, if the Contract Documents are prepared in both languages and there is any apparent discrepancy in the two versions.

Consultant. The architect or engineer engaged by the design-builder is called the ''Consultant.''

Payment Certifier. Due to ethical concerns addressed in **§ 24.16, this supplement,** the contract permits payment to be certified by either the consultant, the owner, or another knowledgeable third party, as designated by the owner. ¶ 14, Document 14. If not certified by the consultant or owner, then a copy of the contract between the owner and payment certifier is to be provided to the design-builder.

Certificates. In wording no doubt requested by the design professionals, there is a statement in ¶ 2.1.3 of Document 14 that certificates issued by the consultant are "to the best of the Consultant's knowledge, information and belief" and are not a guarantee by the consultant that the work is either correct or complete.

Project Mediator. The parties are to appoint a "project mediator" within 30 days after the contract is awarded and all disputes must first go through him or her before arbitration. If arbitration is not demanded by either party within 10 working days after the mediator was requested, then arbitration is not required under the contract.

Insurance and Bonds. Unlike U.S. form contracts, the actual amounts of insurance required are set out in the design-builder-owner agreement, Document 14, including the amount of coverage for errors and omissions and for general liability. For general liability insurance, the policy must be for $2 million per occurrence, with a deductible of not more than $5,000. Professional liability (errors and omissions) insurance limits are less at $250,000 minimum per claim, with an aggregate of $500,000 within any policy year, and must be maintained for two years after substantial completion. Automobile insurance is also set at $2 million. Surety bonds are as agreed by the parties in the latest edition of the CCDC approved bond forms. For Document 15, the design-builder and consultant agreement, it is stated only that the design consultant and its subconsultants shall each carry at least $250,000 per claim with an aggregate limit of at least $500,000 per year and that such insurance must be maintained for two years after substantial performance of the work.

Limitation and Waiver of Owner Claims. Paragraphs 2.1.5 and 12.2.1 of Document 14 attempt to limit the liability of the consultant in actions by the owner. Paragraph 2.1.5 states that the owner waives any right of action in negligence or otherwise against the consultant except that the owner can make a claim against the design-builder under the contract. Paragraph 12.2.1 is an extensive "Waiver of Claims by Owner." The Association of Consulting Engineers of Canada (ACEC) has advised its member that such a clause might not be enforceable since only the parties to the contract are bound by its terms. Here, only the owner and design-builder sign Document 14. However, it is not uncommon for settlement or release agreements to waive claims against parties who are not required to sign that document, as well as their agents, employees, etc. In the United States, such a waiver would be valid even as to non-signing parties. Also, paragraph 1.3 of Document 15 (Design-Builder and Consultant Agreement) incorporates Document 14 by reference, which brings these waivers into the consultant's contract. It is not clear why ACEC would object to a waiver that works in their members' favor. If a consultant wanted to reinforce this waiver, then paragraph 1.1.5 of Document 15 should be amended. It states that, "Nothing in this Contract shall create any contractual relationship between the Consultant and the Owner." If the Consultant is to have the benefit of a waiver signed by the owner, which is a type of contractual relationship, the statement should have added to the end, "*. . . except as to the waivers set out in paragraphs 2.1.5 and 12.2.1 of the Agreement between Owner and Design-Builder.*"

[B] The Association of Consulting Engineers of Canada's Opposition

The Association of Consulting Engineers of Canada (ACEC) took a controversial position in October 2000, when it wrote to its members stating that ACEC does not endorse CCA/RAIC/CSC Design-Build Documents 14 and 15. In fact, they do not refer to either form as a "CCDC" form number. ACEC contends that in working with the other groups to develop these documents, their objectives were not met. ACEC warned its members: "The [ACEC] Board believes that the current Documents create a conflict of interest for the consulting engineer when working as the design-builder's consultant and increase the engineer's liability by leaving an implied duty of care to the owner." ACEC provides its members with a suggested disclosure statement to be given to the owner whenever Documents 14 or 15 are used, stating that the owner understands that:

> [W]hile acting in accordance with its professional obligations, the consultant remains the design-builder's consultant and the owner would not be entitled to rely upon the consultant with respect to any of the design, field review or contract administration services for which the consultant has been retained by the design-builder.

This appears to be an effort to meet Canadian licensing association ethics requirements, but may also be an attempt to avoid direct claims by the owner by disclaiming any intent that the owner rely on the consultant's services. Of primary concern to ACEC is the consultant's role in the preparation of construction cost estimates at early stages of design. Paragraph 2.2.1.7 of Document 15 requires that, during the proposal stage, the design-builder is to "involve the Consultant in the preparation of the construction cost estimates." ACEC's concern is that, at the pre-bid stage, there is no agreement in place outlining the role of the consultant. This is not always true, however, and many design-build teams have preliminary agreements or teaming agreements that address the consultant's role during the pre-bid stage. ACEC fears that as costs escalate during the design development phase, after the design-builder has locked in a contract price, as more information becomes available it will become clear "that the Design-Builder has not budgeted for these additional costs and must absorb them itself." ACEC believes that pressure will be brought to bear upon the design professional to revise the design to bring it into line with the contractor's bid and that this "will inevitably result in creating exposure upon the consultant where the owner sustains losses as a result of the design revisions." (Letter dated July 24, 2000, from ACEC's legal counsel, sent to ACEC members on Oct. 30, 2000). When the costs increase, ACEC fears that the consulting engineer "will be obliged to try to re-design the project to bring it in line with the stipulated price perhaps sacrificing quality in order to do so." ACEC sees this as having two risks: (1) lesser quality in design; or (2) exposure to cost overruns by the contractor. However, it seems that ACEC is reading too much into paragraph 2.2.1.7. Surely the owner expects that in a design-build project the design-builder will "involve the Consultant" when preparing cost estimates. The extent of that involvement can be further detailed by the parties by supplementary conditions and can be defined in the teaming agreement. Nothing in the consultant's scope of work in Document 15 requires it to perform estimating, but only to prepare designs and provide information to the design-builder. A clause could certainly be added that makes clear that the responsibility to prepare cost estimates rests solely with the design-builder and that the consultant relies on the accuracy of those estimates when preparing the design.

ACEC also objects to the requirement that the consultant certify the validity of the punch list and the date of substantial completion. Paragraph 1.5.1.16 requires the consultant to "assist the Design-Builder in determining the date of Substantial Performance of the Work." ACEC warns that the consultant has only carried out periodic field visits and is not in a position to certify the validity of the list. However, this is no different from traditional projects in which the architect or engineer helps prepare the punch list after only periodic site visits and then certifies final completion when the punch list work is done. In either case, the contractor has superior knowledge as to what went on daily and the design professional may not be aware of concealed defects not contained on the punch list. Design-build does not increase the chances of missing things that should be on the list *unless* the design-builder reduces the consultant's frequency of site visits or limits access to the site during construction. Again, ACEC seems to be overly conservative in its reading of the new contract form.

CHAPTER 20

PROJECT MANAGEMENT

§ 20.06 SCHEDULING SKILLS

Page 325, add at end of section:

In a 1999 Nova Scotia court case involving a large design-build power plant project, the court retained its own engineering expert to help advise it on design-build. The court quoted its expert on the challenges of scheduling in design-build and the skills needed. The expert said:

> Design Build requires a large measure of experience on the part of the designers to know, ahead of time, where the Final Design will lead and to make provisions for it at the time the construction is underway. The constructor also requires a large measure of experience to plan the work to prepare for the designs as they are released and have suppliers, subcontractors, and trades ready to supply and do the work. The overall interwoven schedule of design and construction is crucial. The technique of critical path scheduling is vital to this work. Fast Track or Design Build when practiced to the full extent of the art and the skill can significantly shorten the overall construction period. It also allows the Final Design work time to be extended and allows the design group more time to better carry out their tasks without directly impacting the overall project schedule.

Mitsui & Co. Ltd. v. Jones Power Co., 1999 Carswell NS 247 Nova Scotia (1999) Supreme Court.

Inexperience in design and planning, as well as in estimating, can result in cost overruns that give rise to claims and disputes between the design-build team members. *See* **Chapter 23** for some examples of design-build gone bad due to inexperienced personnel being assigned to the project for critical functions.

CHAPTER 21

DESIGN-BUILD RISKS AND LIABILITIES

§ 21.07   STRICT LIABILITY FOR DESIGN-BUILD CONTRACTORS

[E]   Court Cases on Strict Liability in Construction

Page 346, add at end of subsection:

In *Ricci v. Alternative Energy, Inc.*, 211 F.3d 157 (1st Cir. 2000), a worker was killed after he fell 80 feet from a tall smokestack at a power plant in Maine. His father sued the plant owner and the design-build contractor for wrongful death, negligence, and strict liability. The design-builder challenged the strict liability claim, arguing that a power plant is not a ''product'' under Maine law. The trial court granted summary judgment on all claims, but the Court of Appeals reversed that decision and sent the case back to the trial court to resolve the issue of whether a power plant is a product. The outcome of that case is not yet known, but the case shows the potential for a strict liability claim to be made in a design-build setting.

§ 21.11   DEALING WITH MULTIPLE SUBCONTRACTORS

[B]   Subcontractor Delays and Defaults

Page 352, add at end of subsection:

Liability for a subcontractor's defective work can entangle both members of the design-build team, depending on how the team is structured. If the architect and contractor form a joint venture, either one or both may be held liable for damages caused by a subcontractor. One example of such a case is *St. Paul Companies v. Construction Management Co., Ltd.*, 96 F. Supp. 2d 1094 (D. Mont. 2000), in which a fire destroyed a home while it was under construction and nearly complete. The insurer, St. Paul Companies, paid the homeowners $1,123,280 to cover the fire loss and then stepped into the owner's shoes and sued to recover what was paid out under the policy. The defendants included the contractor, the architect, and the electrical subcontractor under claims of breach of contract, negligence, and breach of warranty. Since the evidence showed that the cause of the fire was defective wiring, the subcontractor settled prior to trial and the claims against it were dismissed. The issue at trial was whether the contractor and the architect could be held liable for the subcontractor's negligence. According to the court, the architect and contractor used a "design-build" concept that allowed the customer to deal with only one point of contact for the design and construction of the home. The architect and his firm owned a two-thirds interest in the construction company. The design-build team argued that since the electrician was an independent contractor, not an employee, they could not be held vicariously liable to the owner (or its insurer). Rejecting this argument, the court held that a general contractor, who has a duty to perform a construction contract with due care and in a good and workmanlike manner, could not avoid this duty simply by hiring others to fulfill their duties to the owner. The architect challenged the claim of negligent design, pointing out that the insurer had no expert witness and failed to disclose opinions of an architectural expert. The insurer explained that it was not suing merely for design errors, since the architect "played a broader role in construction of the home, including the design, creation and construction of the home contained in the design-build concept the defendants employed in producing the home." The court found that issues of fact remained for trial and, therefore, denied the architect's motion for summary judgment. Though the eventual outcome of the case is not reported, the message is clear that even when the owner carries insurance to cover damage to the project during construction, there can be liability for work of a subcontractor that causes an insured loss. To avoid such claims, the design-builder should insist on a waiver of subrogation clause in the contract that bars an insurer from suing to recover losses paid out under the owner's insurance policy.

In a 1998 Massachusetts case involving defective work by subcontractors, *Guiliani v. Penny,* 1998 Mass. Super. LEXIS 502, the design-builder hired a structural engineer to design trusses and a framing subcontractor to build them. The owner filed for arbitration when structural defects were found in the trusses, resulting in roof leaks. The design-builder admitted that its two subcontractors were negligent and responsible for the defects. As a result, the court granted summary judgment against the design-builder, stating that as prime contractor, the design-builder was liable regardless of who actually performed the obligations under the prime contract. The court noted: "Under contract law, it is irrelevant whom it [general contractor] directed to perform that obligation, because it remains liable whether the failure was caused by an employee or an independent contractor." The court added that the design-builder may pursue claims against its subcontractors, but that is a separate issue to be tried between them.

CHAPTER 22

JOB SITE SAFETY

§ 22.01 SHIFTING OF TRADITIONAL RESPONSIBILITY

Page 361, add at end of section:

Some design professionals, like structural engineer Emile Troup of Canton, Massachusetts, have concerns about design-build in fast track settings, which he says are really "build-design," where design is on-going while construction moves ahead. Troup says, "Owners think it's great—single source responsibility. But are public safety and the owner's goals well served when the leader of the project is not professionally licensed, not pledged to assure public safety and under the gun to meet an unreasonable budget and schedule?" *Bumpier Road to Finish Line,* Engineering News Record, May 14, 2001, p. 56. Today, every design-build and traditional contract says that "time is of the essence." Fast track, which started in the 1970s, has given way to "flash-track" and, in some cases, "warp-speed" tracked projects. Design professionals need to keep their contractor partners grounded in terms of what can be delivered in a short time frame. Most design errors are made when inadequate time is allowed, or when uncoordinated documents are issued for construction. Faster schedules, and issuance of preliminary documents for pricing and, in some cases, for construction, will naturally lead to oversight and errors. A quality control process must be discussed among the team at the early stages out of concern not only for the completed project but for worker safety.

§ 22.09 LESS LIABILITY THROUGH DESIGNER-LED DESIGN-BUILD?

Page 369, add at end of section:

A case in point is the Kansas Supreme Court decision in *Robinett v. The Haskell Co.,* 12 P.3d 411 (Kan. 2000), in which a design-build contractor was held to be immune from liability to an employee of a subcontractor. Though the subcontractor carried the workers' compensation insurance, the prime design-build contractor was found to be a "statutory employer," immune from liability to the worker just as if the contractor had employed the worker directly. The design-builder was found to have immunity even though the worker alleged that his injuries were caused by the negligence of the design-builder.

CHAPTER 23

WHEN DESIGN GOES BAD: LESSONS LEARNED FOR LITIGATION

§ 23.01 DESIGN AND ESTIMATING ERRORS BY DESIGN SUBCONTRACTOR

Page 388, add at end of section:

In a 1997 Canadian case, the dispute between the teammates on a government design-build project was, again, alleged estimating errors by the design consultant. The result of the case is somewhat surprising. In *Cooper Consultants Ltd. v. Thomas Fuller Construction Co.*, 1997 Carswell Ont 2314 Ontario Court of Justice, General Division, the contractor's estimator was not able to get a steel price from suppliers or subcontractors because the drawings were at only a preliminary stage. He contacted the design engineer for more information so he could calculate the amount of steel needed for the slabs and the vertical elements. A dispute arose after the quantities of steel used turned out to be 50 percent higher than estimated based on advice given by the engineer. The contractor admitted that its estimator made an error in his area calculation of some 1,500 square meters that would result in 32 to 36 times more steel being required than was estimated. The engineer also admitted an error in the information it provided. The parties agreed that there was an overrun of some 550 tons of steel but disagreed on who was responsible. The engineer filed a lien and sued the contractor for $54,965 in unpaid fees. The contractor countersued its teammate, the structural engineer, for $700,000 for negligent misrepresentation. For a claim of negligent misrepresentation, there must be a "special relationship" between the parties. The court found that a special relationship existed here prior to entering into their contract since the engineer assisted the contractor in preparing the bid. However, the evidence showed that the contractor did not rely in a reasonable manner on the misinformation provided by the engineer because the contractor's estimator, an engineering technologist, was inexperienced in estimating structural steel requirements and should have verified the estimates himself, which he failed to do.

The court observed: "In a design build job there [are] no completed drawings with reinforcing steel shown so that they could be given to any subcontractor. The contract dealt with preliminary schematic drawings [which] show openings, the loadings and bay sizes to allow the structural engineer to provide the estimates." The engineer admitted there was a duty under the circumstances for the information that it supplied, but the engineer argued that it was not responsible for the misapplication of the information it furnished. Canadian law, like American law, requires that to recover damages the contractor must show that it "relied, in a reasonable manner, on said negligent misrepresentation" and the reliance must have been detrimental. As a design-build project, there were not finalized plans available during bidding, but only preliminary schematic drawings. The engineer argued that it was the contractor who was negligent and the contractor's own negligence caused the overrun. The court agreed with the engineer, holding that "Fuller [the contractor] was negligent to have someone in this type of project, a design-build, with [the estimator's] limited knowledge and experience." The contractor's claim of $700,000 was dismissed.

Practice Pointer. This case was a close call for the engineer. After admitting a duty to the contractor, as well as erroneous information provided, the engineer took the bold step of blaming the contractor for relying on the information without verifying it. And the court agreed. In the United States, it would not be uncommon in a pure negligence case for the court to have split the damages in some percentage under principles of "comparative negligence," assessing a portion to each negligent party. However, for a negligent misrepresentation claim to succeed, the contractor must show that it reasonably relied on the data. Here, putting an inexperienced estimator on the job was found to be negligence by the contractor. There are lessons here for both parties. First, design professionals must realize that when their teammates ask for either quantity take-offs or cost input during the pricing stages, that information will be used to calculate the contract price. The contractor may merely be asking the architect or engineer for data to compare against the contractor's own estimate, or may be relying entirely on the designer to provide this data. Again, the most qualified member of the team should do each portion and, within each teammate's organization, the most experienced personnel should be assigned to each task. When an apparently inexperienced estimator tells the structural engineer that he does not know how to estimate the steel quantities, a red flag should go up that changes are needed in the staffing. Contract clauses can allocate risk dealing with cost estimating.

§ 23.05   MISREPRESENTATION ABOUT PROJECT SCOPE

Page 392, add at end of section:

A 1997 Canadian case involved a misunderstanding over the scope of work. The project in *Fame Construction Ltd. v. 430863 B.C. Ltd.*, 1998 Carswell BC 2143 (British Columbia S. Ct. 1998), was a medical office building and a pharmacy, plus offices for the provincial Ministry of Health. The design-builder sued the owner over whether "mall standard" specifications applied or whether the design-builder agreed to construct a building fully finished to the owner's specifications. The design-builder alleged that no formal contract was ever executed, and that any written agreement between the parties was not binding. The court found that although a standard form Canadian Construction Association (CCA) 14 contract had been fully executed by the parties, they had put off executing CCA 14 for several months. The contractor delayed signing because it wanted the final construction drawings to be completed to better define the scope of work before it would execute the contract (and lock in its price). The owner (landlord) delayed because it wanted the tenants (doctors) to be legally committed to buying space in the building and did not want to commit to the construction of the building until the tenant was under lease. The court held that the contract here was intended to be a ''design-build'' contract, and that the parties used the phrases ''turn key'' and ''design-build'' synonymously.

The scope of work set out in Article I of the CCA 14, entitled ''The Work,'' read:

> (a) A general description of the work is: The design and construction, in the city of Trail, of a three story Health Centre–Medical Clinic with underground parking.
>
> (b) The Contractor shall, except as otherwise specifically provided, provide all the labour, products, construction machinery, equipment and services required for the performance of all of the Work set out in the Contract Documents and shall forthwith, according to the instructions of the Client, or such other persons as may be designated by him, commence work and diligently perform the respective portions thereof, and deliver the said work to the client within the time specified herein.
>
> (c) The Contractor shall do and fulfill everything indicated by the Contract Documents.
>
> * * * All costs are included in the Turn Key Price, from Design to completion including all and every costs for the completion of this Project including development cost from the City of Trail. * * * The Doctors portion will be designed and built to their exact requirements. All plans and specifications must be approved by the tenants for their portion of space. No extra will be allowed as this is custom design Building to the satisfaction of all tenant specifications unless it has been approved in writing by the owner and the tenant as an extra charge to the tenants.

The court found the phrase ''Total Turn Key'' meant that the entire 46,800 square feet of the building should be constructed on the design-build method and that, by its very nature, agreement on those final drawings could not be an essential element of the contract, stating that ''by agreeing to a design-build contract, [the contractor] had agreed to design and build the entire building, including the doctors' space, to the specifications agreed upon.'' Citing a publication by the Canadian Construction Association (CCA), the court found that,

> The Design-Build Method differs from the more traditional forms of contracting in that the client deals with one single administrative entity, the Prime Contractor, who provides the design and the construction under one Contract Package. During construction, any conflicts between construction and design, and divisions of responsibility between Contract packages, are the responsibility of the Prime Contractor. The Prime Contractor prepares the total design

and thereby retains Architects, Engineers, and other specialists as required; the Prime Contractor is the architect's or Engineer's client.

The court concluded that the misunderstanding arose because the contractor failed to appreciate the difference between a design-build building and a contract to do work according to agreed plans and specifications. The court held that the owner "was entitled to believe that the [contractor], by agreeing to construct a design-built building, had agreed to design and build the entire building." The court found that the contractor had induced the owner into a false belief.

Practice Pointer. This case points out a common dilemma, i.e., the contractor normally wants to wait as long as possible before locking in a price by contract so that designs can be further developed and better cost data obtained to lessen the risk of the unknown. The owner normally wants that price locked in earlier so that it can pursue financing. Another common problem is that the scope of work in many design-build contracts is not well written, leaving ambiguities as to what is intended. In this case, the contractor thought that some minimal level of "warm shell" or "mall standard" specifications was the scope, with tenant finish work in the doctors' offices to be an extra. The owner, however, relied on the term "turn key" and the statement that "[t]he Doctor's portion will be designed and built to their exact requirements" as meaning that the total cost of the build-out of the tenant space was in the contract price, even though the space requirements were apparently not known. This problem could have been addressed by some candid questions during the contract negotiations as to whether the tenant finish work would be handled by change order as an extra, or whether an "allowance" should be included in the contract for that work, with cost adjustments as the tenant's needs and design were further developed. Either way is fair for the owner. However, putting the financial risk of unknown tenant requirements on the design-builder either means that a contingency must be put into the price to cover possible overruns, or the design-builder must absorb the cost of overruns. If the lease agreement includes tenant finish work, then the landlord must give the design-build contractor the cost limits from the lease in order to establish an allowance or working budget. When the tenant's desires exceed the allotted sum, either the parties need to reduce the scope and level of finishes or increase the allowance by change order, billed back either to the owner or to the tenant. The contractor should have a handle on cost sufficient to know when the costs are being exceeded and alert the owner in a timely manner.

§ 23.10   UNCLEAR PAYMENT TERMS IN PRIME AND SUBCONTRACTS

Page 398, add at end of section:

In this changing industry, design professionals wear many hats, including that of prime design-build contractor in some cases. When design professionals solicit bids from trade subcontractors, they must be clear whether they are doing so on the owner's behalf, such as a construction manager, or on their own behalf, as a design-build contractor. Confusion as to which hat is being worn can result in litigation. One such lawsuit is a Canadian case, *Wm. Roberts Electrical & Mechanical Ltd. v. Kleinfeldt Consultants Ltd.*, 1985 Carswell Ont 958 (Ontario Supreme Court, High Court of Justice 1985), in which a mechanical/electrical subcontractor sued a firm of consulting engineers for $27,000 after the project owner went bankrupt without paying for construction work. The engineering firm had been hired by a manufacturing company to prepare plans for a ventilation system in the plant printing room, to seek tenders (bids), to make recommendations, to supervise the work, and to approve the payment of the subcontractors' invoices. The engineer prepared a drawing for the work and called six subcontractors for bids. The letter enclosing the drawings invited the bidders to submit a price for the work. The successful bidder sued the engineer for payment after the owner went bankrupt, contending that the engineer was actually a design-build contractor and was, therefore, responsible for payment. The case involved expert testimony of two professional engineers, who both agreed that consulting engineers did on occasion act as design-builders, in which case they functioned as the general contractor and, therefore, were responsible for payment of the subcontractors. The plaintiff had had some experience in such contracts with engineers as prime. All . the parties admitted that such a contract was unusual, however. The plaintiff's own expert engineer stated that such a method would not be used on a $26,000 job. The court ruled in favor of the engineering firm, stating:

> [I]t should have been clear to any reasonable businessman, and clear to the plaintiff from the letter of September 24 and the enclosed plan, that the defendants were consulting engineers and nothing else, and that they were inviting bids on behalf of [the owner] for work to be done by them on their credit and not on the credit of [the engineering firm] . . . I also find that a design-build arrangement is a relatively unusual one and if the plaintiff was not certain about who the paymaster was, he had an obligation to inquire.

Practice Pointer. There are lessons here for design firms and for trade subcontractors. In this new world of changing roles, with architects becoming contractors in design-build, design professionals must realize that there can be honest confusion. Suppose an engineering firm is the prime design-build contractor on Project A, soliciting bids for its own use in preparing a bid to the owner, while on Project B that same firm serves as construction manager, or merely obtains bids to pass on to the owner for a construction project. A subcontractor bidding on both jobs might need some clarification as to whom it is bidding to and for what purpose. The ultimate question is: Who is going to pay me? Here, the court found that the information in the request for bids was, or should have been, clear to any reasonable trade subcontractor as to the engineer's role. But the need for clarity will only increase as design firms slip into the role of general contractor in design-build. Subcontractors need to ask questions they never thought of before to be sure they know who is going to pay them. The case also points out the problem of owner bankruptcy, when anyone who is left with an unpaid bill will look to someone with money to pay it. Clever legal arguments can attempt to turn a solicitation for one purpose into something totally different in an effort to find a solvent person to pay the bill. A sentence or two in the RFP (or tender solicitation) could save legal fees later on.

§ 23.12   ACTION FOR FRAUD BASED ON MISREPRESENTATION

Page 401, add new subsection at end of chapter:

[C]   Action for Fraud Based on Scope Change by Owner

Though the trial court opinion does not resolve the case, the facts in *ABB Daimler-Benz Transportation, Inc. v. National Railroad Passenger Corp.*, 14 F. Supp. 2d 75 (D.D.C. 1998) indicate that a major dispute arose when the owner made a scope change during the design process. The plaintiff, Adtranz, designs, manufactures, and sells electrical equipment for rail transportation systems. Defendant New Jersey Transit (NJT) provides commuter passenger rail service over tracks controlled by defendant Amtrak. NJT and Amtrak formed a joint venture as owner to award a design-build contract to increase the available power on Amtrak's traction power system by converting 60 Hz electric energy from power lines into 25 Hz power, suitable for use by railway traction equipment. In response to Amtrak's Request for Proposals (RFP) for the project, Adtranz entered into a teaming agreement with L.K. Comstock & Co., Inc. (Comstock) to prepare and submit a proposal for the project, jointly signing a contract with Amtrak, and jointly performing any contract they were awarded. Under the teaming agreement, Adtranz took responsibility for the design and procurement of equipment and Comstock took responsibility for all construction. After Amtrak awarded the design-build contract to Adtranz, Amtrak rejected the design-builder's 60 percent Design Submittal and required that the design be changed to a different technology at no additional cost to Amtrak. This change in technology significantly increased costs for both Adtranz and Comstock by approximately $10 million. Adtranz alleged that rejection of the 60 percent Design Submittal was based on the setting of new performance requirements that were inconsistent with those set out in the original RFP and the contract. Adtranz further alleged that the change in performance requirements was part of a conspiracy by NJT and Amtrak to force Adtranz to upgrade the original project requirements to the more expensive technology, at no additional cost to Amtrak. Adtranz further claimed that the changed design specifications caused Adtranz to incur significant cost increases and other losses as a result of being forced to redesign the project. Adtranz sued both NJT and Amtrak for breach of contract, fraud, *quantum meruit,* and unjust enrichment. Design-build partner Comstock joined in the suit and sought damages for increased construction costs, claiming that Amtrak's wrongfully imposed change in technology to dc-link increased construction costs. Comstock also asserted a fraud claim against both NJT and Amtrak.

Comstock argued that it had the right to sue Amtrak under the teaming agreement, which divided responsibility for performance of the contract between the two companies. Comstock claimed that the agreement gave it the right to sue Amtrak in the name of Adtranz for any claims related to its construction of the Project. The court disagreed, finding that the owner did not sign the proposed teaming agreement and, therefore, was not bound by it. As to the claims by Adtranz, New Jersey law precluded Adtranz's claims for fraud, conspiracy to defraud, and a tortious interference against NJT. Therefore, the court dismissed those claims as well as the claim for punitive damages. Regarding claims against NJT, New Jersey law bars claims for contracts implied in law as well as those based upon implied warranties. As a result, the court denied the claims for unjust enrichment and *quantum meruit* as well.

Practice Pointer. One of the few bases for a change order in design-build is a scope change by the owner. That is why it is important to have a clear scope in the contract so that a change is easy to detect. In this case, the change in technology appears to have been a change in what was represented in the original RFP and the contract. However, the scope may have been vague enough to allow the owner to contend that the change was within the original scope. The performance criteria in the RFP should leave discretion to the contractor to determine the best way of achieving the owner's goals, at the least expense. The design-build contract should also include dispute resolution procedures when there is a question over scope changes made by the owner. In many cases, such disputes are negotiated. However, on a large project like this one, where a scope change results in $10 million in added costs, parties tend to resort to more traditional means of litigation to work things out. A better-defined scope and recognition by the owner of how changes in technology can impact costs may have paved the road to a smoother project for this team. It is important to spend time up front to develop a detailed written scope of work and incorporate that into the prime design-build contract. If the scope and programming are part of the design-builder's tasks, then at least make reference to a scope "later to be developed and agreed to in writing by the parties"—and then do it.

CHAPTER 24

ETHICAL AND CRIMINAL ISSUES IN DESIGN-BUILD AND CONSTRUCTION

§ 24.01 CONFLICTS OF INTEREST

Page 407, add new subsection after [D]:

[E] Adverse Effects of Conflicts

Conflicts of interest are not just an issue for design firms. When a design firm does repeat business with a contractor teammate and then has the opportunity to recommend that contractor as a qualified bidder to an owner on another project, the design firm's neutrality may be called into question. The result can be the disqualification of the contractor teammate for a traditional project involving the same design firm. That was the situation in an Illinois case, *Joseph J. Henderson & Son, Inc. v. City of Crystal Lake,* 743 N.E.2d 713 (Ill. App. 2d Dist. 2001), which involved a bid protest on a wastewater treatment plant. The city solicited bids to improve a wastewater treatment plant. Only two bids were submitted, Henderson's bid of $13.9 million, and Seagren's bid at $14.2 million. The city council rejected Henderson's low bid because of concerns that Henderson's business relationship with the project's engineer would ''create the appearance of impropriety.'' Henderson sued asking that the court order the city to award the contract to Henderson. The city's engineering firm, Baxter & Woodman, Inc., also owned 79 percent of a limited liability corporation named B & W Design/Build L.L.C. (B & W). For several years, B & W had contracted with Henderson to provide design and construction supervision services on numerous projects. All of B & W's profits for 1999 came from work B & W had done jointly with Henderson. The city's mayor testified that he and the city council were worried that the relationship between Henderson and B & W would create an appearance of impropriety. The trial court ruled that the city's award of the contract to the next higher bidder was not based on favoritism, unfair dealing, fraud, or any other arbitrary conduct. The Court of Appeals agreed that the award of the contract to the higher bidder was an exercise of the city's discretionary power done in the public interest, stating: ''Maintaining the public's confidence in a multimillion dollar project funded with public monies was a legitimate consideration for the city council when determining the lowest responsible bidder.'' The bid protest was rejected.

§ 24.02　CONSENTING TO ILLEGAL AND CODE VIOLATIONS OR ACTIONS AGAINST ADVICE

Page 408, add at end of section:

Like AIA Rule 2.105, Bylaw 32.3 of the Architectural Institute of British Columbia (AIBC) requires that architects who become aware of an action taken by the architect's employer or client, against the architect's advice, that violates legal requirements, must take all reasonable steps to convince the employer or client to comply with the law and (1) refuse to consent to the action, and if not rectified in a timely manner, then (2) report the action to the authority having jurisdiction and, if the authority confirms the violation is not rectified in a timely manner, then (3) terminate services on the project. Official commentary to the British Columbia Bylaws states that "legal requirements" includes building laws and regulations, including health, zoning, development permit, and building permit requirements. The comments also note that termination "is a last resort in the face of an action taken and persisted with by an employer or a client" after the architect has exhausted all available appeals.

§ 24.03　MAKING DECISIONS ON PAY APPLICATIONS, COMPLETION DATES, OR CLAIMS

Page 411, add new subsection after [B]:

[C]　The Canadian Perspective

Design professionals practicing in Canada recognize the potential conflict of interest in certifying a design-builder's pay applications. The Association of Consulting Engineers of Canada (ACEC) has warned its members about conflicts present in the approval of progress payments to the design-builder that places the consultant in a difficult position. Certification of the design-builder's application for payment may include the design firm's own invoice or may relate to some savings the design firm will share with the contractor. This raises a potential that decisions will be made based on what is most profitable for the design firm, rather than what is in the owner's best interest. Canada's standard form design-build contract (CCA/CSC/RAIC Document 15) places the design professional (the "consultant") in a subcontractor role to the design-builder. Paragraph 1.5.1.8 requires the consultant to determine the value of work performed and products delivered in support of the design-builder's application for payment. While the consultant has no direct contract with the owner, Canada's ethical rules would require a disclosure to all persons that might be affected by an apparent or potential conflict like this. According to ACEC, the consultant is expected to be totally impartial, and owes a duty to the owner to be neutral with respect to certifying payment. However, as a subcontractor to the design-builder, ACEC says that the contractor "controls the purse strings" and might withhold payment from the consultant in order to force the consultant to over-certify payment. This fear of economic pressure, however, is no different or greater than that pressure placed on design professionals by owners in traditional delivery systems, demanding that the consultant withhold payment to the contractor due to delay, disputed extras, or defective work. Owners also "control the purse strings" and frequently try to insert their judgment on issues of aesthetics, acceptable work, weather delays, and owner-caused delays, and press their architect or engineer to take positions siding with the owner against the contractor. Consultants are either able to stand up to their client or they are not, whether in a traditional role or in a design-build relationship. The consultant's ethical obligation is no less when working for a contractor than when working for an owner. Both can exert economic pressure on a design professional. Ethical consultants will do the right thing under either form of delivery system.

More troublesome is the ethical dilemma when the design-builder's pay application includes *the consultant's own invoice*. Can the consultant impartially approve his own invoice and recommend payment to the owner? While it is apparent that the consultant believes it is owed the amount in its invoice, some would suggest that the consultant should either: (1) certify all but its own invoice; or (2) make a written disclosure to the owner of what must be an already obvious situation. Canada's standard CCA/CSC/RAIC Document 14 gives the owner the option to hire a payment certifier other than the consultant, so the owner can bring a neutral party into that process if it has any concerns about the consultant's loyalties. For those owners who need the assurance of a true neutral, they can hire a payment certifier to handle that task. It is interesting that design-build has spawned these new job positions such as "bridge consultant" and now "payment certifier." How long until they have their own professional associations?

ACEC also warns its members that, under Document 14, the consultant is to certify payment based on "the agreed schedule of values," indicating that the consultant must accept the design-builder's schedule for purposes of payment certification. If the design-builder is involved in "front end loading" of its schedule of values in order to draw off more profit up front, does the consultant have to object, or must it accept the design-builder's schedule of values? It seems that the answer lies in the fact that Document 14 states in paragraph 5.2.3 that this schedule of values is submitted to the owner early on, "supported by such evidence as accepted by the Owner." Acceptance by the owner means that the owner and design-builder have agreed that the schedule is reasonable. If these two agree, this would seem to remove any ethical concerns by the consultant, whose job is then to merely apply that schedule to the percent of work satisfactorily performed.

§ 24.12   GOVERNING LAWS

Page 426, add new subsection after [H]:

[I]   Canadian Price Lists and Tariffs

In the United States, continuing into the 1960s, it was thought by most architects that the Sherman Antitrust Law did not apply to professionals and, as a result, the AIA's code of ethics prohibited its members from competing on the basis of fees. One AIA ethical standard required members to use the AIA chapter fee schedules to set rates. In later years, activists pressed the Justice Department to apply antitrust laws to professionals and, eventually, the AIA and others repealed any fee schedules. Carl M. Sapers, Esq., *Ruminations On Architectural Practice,* Constr. Contracts Law Report, Vol. 25, No. 8, April 20, 2001, p. 3. However, in other countries fee schedules are still used. For example, Canadian provinces still publish provincial architectural and engineering scales and tariffs for professional fees on various types of projects. However, it is reported that Canada's licensing associations have begun to question the legality of such fee schedules and do not enforce their requirements.

The licensing laws in Ontario and British Columbia contain a Tariff of Fees for Architectural Services. The Architectural Institute of British Columbia (AIBC) Tariff is available at ⟨www.aibc.bc.ca/public/seeking_an_arch/tariff.html⟩ and includes minimum recommended hourly fees for principals and staff and a Minimum Net Percentage Fee Scale for Basic Architectural Services for various project types, with adjustments for new construction and renovations. The suggested fees range from a minimum of 2.3 to 25 percent, depending on project type and size. The AIBC's Bylaws state that, except when providing *pro bono* services or as approved by the licensing board (the Council), "an architect shall provide services and receive fees in substantial accord with the Tariff of Fees for Architectural Services." The Tariff of Fees is authorized by statute in British Columbia. Section 24(2)(e) of the Architects Act of British Columbia. Commentary to the Bylaw explains that it has been "the profession's long standing policy" that the compensation for services should be per the published Tariff of Fees, which has a "long and tested tradition" of being an equitable level of compensation. The Commentary explains that the Tariff of Fees is a general guidance for appropriate fees, neither a price list nor a minimum or maximum. The Tariff of Fees is published as a budgetary check so that an architect might know when fee levels are too low to provide adequate levels of service. The term "substantial accord" in the Bylaw shows that this is similar to a manufacturer's "suggested retail price"—not required, but suggested.

Page 427, add new sections at end of chapter:

§ 24.16   ETHICAL ISSUES IN CANADA

There are many similarities in the Codes of Ethics of the licensing boards in the United States and those in Canada. The Canadian Design-Build Institute (CDBI) *Practice Manual* notes that design-build has raised ethical and professional questions for architects and engineers. CDBI reports that in 1996 the Royal Architectural Institute of Canada surveyed provincial architectural associations about professional practice questions related to design-build. As in the United States, the survey found that regulations across Canada varied and were frequently silent or unclear about design-build and the related professional and ethical issues.

[A] Conflicts of Interest

Canadian licensing boards and professional associations have all included conflicts of interest in their Codes of Ethics. For example, in Alberta, the Code of Ethics states: "An architect should act impartially and not place himself in a position of a conflict of interest." Rule 5, Code of Ethics, Alberta Regulations 240/81. Disclosure of conflicts of interest is a uniform ethical obligation for architects and engineers practicing in Canada. The Alberta Code continues that once a conflict arises, "it should be immediately disclosed by the architect to his client personally." Another typical disclosure requirement is in the Code adopted by the Association of Consulting Engineers of Canada, which states quite simply: "Members shall disclose any conflicts of interest to their clients." The Code of the Canadian Council of Professional Engineers is similar. Other professional codes in Canada go into more detail, such as the Canons of Ethics of the Nova Scotia Association of Architects, which states: "An architect shall not assume or consciously accept a position in which his interest is in conflict with his professional duty."

Similarly, the Code of Ethics and Professional Conduct for the Architectural Institute of British Columbia (AIBC), Bylaw 31, Conflict of Interest, states:

> Except as permitted hereunder and with full disclosure under Bylaw 32.0, an architect shall avoid actions and situations where the architect's personal interest conflict or appear to conflict with professional obligations to the public, the client and to other architects.

AIBC's Code of Ethics Bylaw 31.2 states:

> An architect having a personal association or interest, which relates to a project, shall fully disclose in writing the nature of the association or interest to the architect's client or employer. If the client or employer objects, then the architect will either terminate such association or interest or offer to give up the commission or employment.

Commentary to this Bylaw explains that "personal" interest includes direct or indirect potential for financial or material gain. If acting as both designer and builder there is an obvious direct potential for financial gain. If acting as a subcontractor or joint venture partner to a contractor, the same potential exists. A literal reading of the Bylaw requires the architect to state the obvious fact that the architect stands to gain financially through design-build and obtain the client's consent.

One example of conflicts of interest is allegedly present in standard Canadian design-build form contracts. The Association of Consulting Engineers of Canada (ACEC) has openly opposed the 2000 edition of CCA/RAIC/CSC Design-Build Documents 14 and 15, in part due to perceived conflicts of interest. In a letter to their members in October 2000, ACEC warned that Document 15 requires the consultant to be the interpreter of the contract documents and to make findings on claims, disputes, and other matters in question relating to the performance of the work or the interpretation of the contract documents. ACEC advises its members using these forms to warn the owner that it should retain its own advisor with regard to field services or contract administration. This seems to go against the streamlined concept of design-build. It adds another design firm to the project costs, one that might be influenced by the owner rather than the design-builder. ACEC points out that under Documents 14 and 15 the consultant is required to report defective work only to the design-builder, but the owner is the only one who can insist that defects be corrected. As a result, ACEC fears that the consultant's reports will go no further than the design-builder, who will not share them with the owner, thus placing the consultant in the position of knowing about defects while the owner does not. However, contracts are not the only source of an architect's duties. There are current ethical rules in some provinces to address this very situation. *See* **§ 24.02,** *supra* **this supplement,** and the ethical rules that require architects to tell the appropriate official if their employer or client takes action contrary to laws that affect public safety. Thus, the architect who has knowledge of a code violation must not keep silent, and must report such action even though reports of defective work are given only to the design-builder by contract.

[B]   Providing Free Services

As late as 1977, the AIA's Code of Ethics prohibited its members from offering or providing ''free design sketches, models or other architectural services, except through design competitions.'' Rule 209, AIA Code of Ethics and Professional Conduct, AIA Doc. J330 (1977). That rule has long since been repealed and the AIA is prohibited by a Consent Decree with the United States Department of Justice from discouraging its members from providing discounted or free services. Final Judgment, *United States v. The American Institute of Architects,* Civil Action No. 90-1567 (Oct. 31, 1990). Some licensing board rules in Canada still prohibit architects from providing free services, however. For example, Bylaw 34.10 of the Architectural Institute of British Columbia states that, except in an approved competition, architects shall not provide services until retained by the client. Official commentary explains that this bars ''speculative services to lure or entice a client,'' and that, prior to being retained, an architect is not permitted to provide solutions, suggestions, or ideas in any format that have value to the client. This would effectively prohibit design-build competitions where the architect is required to prepare and submit design solutions as part of the technical proposal unless there is an approved competition. The Commentary goes on to state that this rule applies specifically to ''design, costing and technical matters.''

The AIBC Bylaws permit ''contingent'' fee arrangements, however, where an architect prepares preliminary design work to assist a client but only when risk of financial failure is high. Commentary on this Bylaw explains that, in such high risk projects, the contingent fee ''shall be no less than three times the fees as described in the Tariff of Fees for Architectural Services.'' The apparent message in this province is that architects should charge their design-build partner a fee to prepare preliminary documents that are part of the technical proposal in a design-build competition. If a firm is giving away its services, at risk, the fee upon selection of the team should be three times a normal fee, according to AIBC's Bylaws.

[C] Sharing Stipends

Bylaw 34.15 of the Architectural Institute of British Columbia states: ''An architect receiving monies for services provided by others shall not use such monies for the architect's own purposes, and shall distribute them promptly to those so entitled.'' If a teaming agreement sets out how a stipend is to be shared, it would not only be a breach of that contract, but an ethical violation for the architect not to share the stipend with its contractor teammate and, if previously agreed, with other design consultants. It is recommended that in a teaming agreement there should be some understanding on how a stipend will be shared by the team participants.

[D] Compensation from More than One Client

Ethical rules might come into play in a situation where a design firm has been hired as the owner's consultant to prepare design criteria (the bridge consultant) and then attempts to compete for the work as part of a design-build team, or where the project starts off as a traditional project but, sometime after the design phase begins, the owner changes direction and wants to use design-build. Can the original design firm become part of a design-build team? Can that firm also stay on as the owner's consultant? In either situation, the design firm may receive payment from more than one client during the course of the project, first from the owner and, later, from the contractor. The AIBC Code of Ethics Bylaw 31.1 states: ''An architect shall not accept compensation for services from more than one party . . .''

Likewise, the Canons of Ethics of the Nova Scotia Association of Architects states: ''An architect shall not accept compensation, financial or otherwise, from more than one interested party for the same services, or for services pertaining to the same work, without the consent of all interested parties.''

Disclosure is the key to avoiding an ethics complaint, since the Canadian boards seem to accept a written disclosure signed by the owner when there is a conflict like this.

[E] Compensation Derived from Construction Profits

The old AIA Code of Ethics prior to 1978 prohibited AIA members from engaging ''in building contracting where compensation, direct or indirect, is derived from profit on labor and materials furnished in the building process.'' *See*

§ 24.01[A], main volume. While not as specific in its wording, consider the Code of Consulting Engineering Practice adopted by the ACEC, which states: ''Members shall obtain remuneration for their professional services solely through fees commensurate with the services rendered.'' How does this affect an engineer who partners with a contractor in a design-build venture in which the two firms share profits from the construction? Can the engineer obtain payment by sharing in the savings on the construction cost? What if there is a large bonus for early completion? Is the engineer's fee limited to only the reasonable value of the design work, or can an engineer make a windfall profit for a job completed ahead of schedule? These questions remain unanswered, but the safest approach in Canada is to disclose any conflict (including a shared savings or bonus) and get the owner to sign it.

[F] Design Competitions Restricted

In Canada, design competition among consultants for the same project, or same client, is not permitted unless by an approved competition. For example, the Canons of Ethics of the Nova Scotia Association of Architects state:

> An architect shall not take part in an architectural competition which is not carried out in accordance with the Royal Architectural Institute of Canada, Code for the Conduct of Architectural Competitions approved by the Council of the Association and in force at the time of the competition nor shall he take part in any competition which the Council has declared to be irregular.

The Canadian Design-Build Institute (CDBI) states that the ethical restriction on competitions ''applies to design-build proposal calls, in that, without an approved competition in place, no more than one consultant is permitted to provide ideas to the same proponent design-builder or to the owner.'' CDBI notes, however, that competition through design-build proposal calls is currently under review by provincial associations across the country. Watch for changes in the By-laws and regulations of the various licensing associations on this topic.

[G] Competition on Fees

Competitive bidding for architectural services is prohibited in most states by the Brooks Act and the state mini-Brooks Acts. *See* **§ 14.05, main volume.** However, fee competition in the private sector is not barred and, in fact, the American Institute of Architects (AIA) is prohibited by a Consent Decree with the U.S. Department of Justice from directly or indirectly adopting any rule, policy, or statement that is unethical to compete based on competitive bids for fees. Final Judgment, *United States v. The American Institute of Architects,* Civil Action No. 90-1567 (Oct. 31, 1990). However, Canadians have quite a different situation. For example, the Canons of Ethics of the Nova Scotia Association of Architects state: ''An architect shall not compete with another architect on the basis of charges for work.'' This would prohibit competitive bidding among design firms seeking to team up with a contractor on a design-build project. The rule suggests that design-builders in Nova Scotia must hire the most qualified architect or firm based on qualifications rather than price.

§ 24.17   INTERFERENCE WITH CONTRACT

From at least 1963 to 1975, the AIA's Code of Ethics prohibited its members from soliciting work for which another architect had already been employed. Standard 9 of the May 1975 Code stated:

> An architect shall not attempt to obtain, offer to undertake or accept a commission for which the architect knows another legally qualified individual or firm has been selected or employed, until the architect has evidence that the latter's agreement has been terminated and the architect gives the latter written notice that the architect is so doing.

However, this rule was challenged as anti-competitive and was dropped after the AIA was sued by a member for antitrust law violations. Although no longer part of the AIA's Code, the rule against pursuing projects already under contract to another is alive and well in parts of Canada. For example, the Canons of Ethics of the Nova Scotia Association of Architects state that:

> An architect shall not undertake a commission for which he knows another architect has been employed until he has notified the other architect of the fact in writing and has conclusively determined that the original employment has been terminated.

A similar rule is contained in Bylaw 34.7 of the Architectural Institute of British Columbia, which states: ''An architect shall not supplant or attempt to supplant another architect after the other architect has been retained or definite steps have been taken toward the other architect's retention.'' Official commentary to the AIBC Bylaw states that a client is free to dismiss the architect at any time and the intent of the Bylaw is not to ''protect either mismatched clients or architects.'' However, the Bylaw is aimed at protecting the client/architect relationship, particularly in the early stages, when mutual trust and understanding are being formed. The companion Bylaw 34.8 bars an architect from accepting a commission for a project until the services of any architect previously retained for the project have been terminated. Therefore, if a design-build contractor seeks to replace its architect, the replacement firm needs to check state or provincial licensing laws before undertaking such work.

Page 448, after Appendix B add new appendix:

APPENDIX B1

SUPPLEMENTARY CONDITIONS TO AIA DOCUMENT NO. B901 (1996 EDITION); STANDARD FORM OF AGREEMENT BETWEEN DESIGN-BUILDER AND ARCHITECT

[Note to user. These Supplementary Conditions should be considered by architects who are asked to act as subcontractors to general contractors under the American Institute of Architects (AIA) design-build contract forms.]

These Supplementary Conditions amend AIA Doc. No. B901 published by the American Institute of Architects (AIA), titled Standard Form of Agreement Between Design-Builder and Architect, 1996 Edition.

SC-1 Standard of Care. Architect shall perform its services in accordance with the standard of care for architects in the state in which the Project is located, performing similar services under the same or similar circumstances.

SC-2 Estimating Errors. In performing design services, Architect is expected and entitled to rely on Design-Builder's estimating services. The Architect shall not be required to revise the design or the Contract Documents at Architect's own cost where such revisions are due to estimating errors, mistakes or miscalculations of market conditions by Design-Builder. The Design-Builder shall pay Architect as Additional Services the cost of design revisions needed to the extent caused by such estimating errors.

SC-3 Site Observations. Design-Builder recognizes that absent full-time resident services by Architect, much of the Work will go unobserved by the Architect and that full-time resident services are not part of Architect's contract services. Endeavoring to guard against defects and deficiencies in the construction does not make the Architect responsible for Design-Builder's errors or omissions.

SC-4 Certifications. With regard to any certifications Architect is requested to sign, at a minimum, the Architect has the right to: (1) approve the form of certificate in advance; and (2) issue only those certificates Architect can reasonably do based on its level of involvement during construction. No such certification shall require Architect to issue any warranty or guarantee of the design or construction work.

SC-5 Ownership and Copyrights in Work Product.   Delete paragraph 3.3 and re-place with the following new subparagraphs:

1. Upon receipt by Architect of final payment, the Design-Builder may provide the Owner with a set of reproducible documents for Owner's use and reference in occupancy of the Project, from which the Architect's name, logo and seal shall be removed.

2. The copyrights in Architect's drawings, specifications and other documents ("Work Product"), including any Architectural Works, shall remain with the Architect. Notwithstanding any provision of the Prime Agreement, both Design-Builder and Owner are prohibited from future reuse of the Work Product on this Project or another project without an agreement in writing acceptable to Architect.

3. In the event that either the Owner or Design-Builder desires to use the designs or Work Product of the Architect for other projects or for completion of this Project after termination of this Agreement, the Architect and such parties may agree in writing to a limited license for such use provided: (1) Architect is first paid in full for all services rendered and expenses incurred in connection with this Agreement; (2) the party seeking Architect's consent agrees that such use without Architect's involvement is at the user's sole risk and without liability to Architect and further agrees to release, indemnify, defend and hold harmless Architect, its subconsultants and their respective officers, directors, employees, and agents from any claims, damages, losses, liabilities, cost or expense, including attorneys' and expert fees, whatsoever arising in any manner from such use including (where permitted by law) those claims alleging the negligence of any such indemnified person or party; and (3) Architect is paid an additional sum of $______________. This paragraph shall survive any termination of the Design-Build Agreement or this Agreement.

SC-6 Disputes.   No decision by Owner or result of proceedings between Design-Builder and Owner shall be conclusive on Design-Builder's liability to Architect for any time extension or additional compensation or other relief.

SC-7 Payment.   Design-Builder shall make payment to Architect within seven (7) days after receipt of payment from Owner, but in any event not later than forty-five (45) days after the date of Architect's invoice. Nothing in the Agreement is to be interpreted as contingent "pay when paid" or "pay if paid" clauses. No payment to Architect hereunder is conditioned on the Owner having first paid the Design-Builder. Interest on amounts due and unpaid shall accrue at the rate of 1.5% per month or the highest legal rate permitted by law, whichever is less.

SC-8 Insurance.
1. The parties recognize that Professional Liability Insurance is typically written on a one-year, claims made basis and that Architect cannot guarantee that such insurance will be available at reasonable rates after Substantial Completion. Design-Builder shall not insist on a certificate of insurance beyond that which Architect can reasonably provide. In addition, nothing in this Agreement shall require the Architect to name the Design-Builder, Owner or others as additional insureds on Architect's

Professional Liability and Workers Compensation policies, which normally do not permit additional insureds.

2. Design-Builder shall name the Architect as an additional insured on the Design/Builder's Commercial General Liability policy and Comprehensive Automobile Liability policy of insurance and shall provide the Architect with a certificate of insurance evidencing such coverage.

3. In paragraph 7.4, replace the reference to "AIA Document A201, General Conditions of the Contract for Construction" to "AIA Document A191, Standard Form of Agreement Between Owner and Design-Builder."

SC-9 Conflicts.   In the event of a conflict between the provisions of this Agreement and the prime agreement, the provisions of this Agreement shall control.

SC-10 Release of Claims.   Nothing in this Agreement shall be construed as a release of claims by Architect that first arise after the submission of the Architect's final Application for Payment. Acceptance of final payment by Architect shall not be a waiver of any claims previously made in writing and identified as unsettled at the time of Architect's final Application for Payment.

SC-11 Indemnity.
A. Subject to the limitation set out in SC-23, below, Architect agrees to indemnify and hold harmless the Owner and Design-Builder, their respective directors, officers, and employees from and against all losses, damages, costs and expenses, including reasonable attorneys' fees, but only to the extent caused by the negligent acts, errors, or omissions of the Architect, its consultants and employees and those for whom they are liable.

B. Design-Builder agrees to indemnify and hold harmless the Architect, its consultants and their respective directors, officers, and employees from and against all losses, damages, costs, and expenses, including reasonable attorney's fees but only to the extent caused by the negligent acts, errors, or omissions of the Design-Builder, its subcontractors (other than Architect), suppliers, and employees and those for whom they are liable.

SC-12 Safety Responsibility.   Nothing in this Agreement shall be construed as making Architect a Controlling Employer as defined by OSHA for purposes of site safety.

SC-13 Licensing.   The Design-Builder is duly qualified, licensed (where required), and authorized by law to act as general contractor and to offer design-/build services in the state where the Project is located.

SC-14 Severability.   If the contract between the Owner and Design-Builder is incorporated by reference into this Agreement, then to the extent that the prime agreement is declared invalid or unenforceable, this Agreement shall be deemed severable. Any provision or part of the prime agreement that is held to be void or unenforceable under any law, shall be deemed stricken and all remaining provisions of this Agreement shall continue to be valid and binding upon the parties, including the Architect's right to payment for professional services rendered and expenses incurred. Architect's right to payment is not contingent upon the validity of the prime agreement.

SC-15 Electronic Media. Design-Builder may request computer disks containing some or all of the work prepared by the Architect either as record drawings or for use by subcontractors in shop drawings. Due to the potential that the information set forth on electronic media (disk) can be modified by the Design-Builder, or others, unintentionally or otherwise, and that design changes may be made after the data is downloaded onto a disk, the Architect shall remove all references to its authorship, corporation name, and/or involvement from each electronic display. Design-Builder recognizes that use of such electronic media (disk) will be at Design-Builder's sole risk and without any liability, risk, or legal exposure to the Architect, its employees, officers, or consultants. The parties shall execute a separate agreement setting out the rights and responsibilities relating to electronic documents requested of Architect. [**See Appendix D1 to this Supplement.**]

SC-16 Record Drawings. If Architect's Basic Services include the preparation of a set of reproducible record drawings showing significant changes in the Work made during construction, then Architect may utilize marked-up prints, drawings, and other data furnished by the Design-Builder upon which the Architect shall be entitled to rely.

SC-17 Additional Services. Basic Services of the Architect include the following. Services beyond these limits shall be provided as Additional Services in accordance with Paragraph 9.2.1:

.1 up to ____ (__) reviews of each Shop Drawing, Product Data item, Sample and similar submittal of the Design-Builder.

.2 up to ____ (__) visits to the site by the Architect over the duration of the Project during construction.

.3 up to ____ (__) inspections for any portion of the Work to determine whether such portion of the Work is substantially complete in accordance with the requirements of the Contract Documents.

.4 up to ____ (__) inspections for any portion of the Work to determine final completion. The following services shall be deemed Additional Services:

.5 review of a Design-Builder's submittal out of sequence from the submittal schedule agreed to by the Architect.

.6 responses to Design-Builder's requests for information where such information is available to the Design-Builder from a careful study and comparison of the Contract Documents, field conditions, other Owner-provided information, Design-Builder-prepared coordination drawings, or prior Project correspondence or documentation.

.7 Change Orders and Construction Change Directives requiring evaluation of proposals, including the preparation or revision of Architect's drawings or specifications.

.8 providing consultation concerning replacement of Work resulting from fire or other cause during construction.

.9 providing services in preparing or filing documents for storm water runoff permits.

SC-18 Mediation. The mediation shall be held in *[insert selected city and state]* unless another location is mutually agreed upon by Design-Builder and Architect.

SC-19 Termination. Add the following new Paragraph 8.9: "Architect shall have no liability for drawings, documents, designs or services that are incomplete due to an early termination of this Agreement. The Owner and Design-Builder assume the risk of errors or omissions in drawings it uses that are incomplete due to termination of this Agreement. The parties also recognize that drawings prepared for design-build projects may not be fully developed and, therefore, are not suitable for use by other contractors without Architect's involvement."

SC-20 Payment Bonds. If Design-Builder provides Owner with a payment bond for this Project, then Architect's services shall be considered to be labor and Architect's Work Product shall be considered to be material for purposes of such bond and Architect shall be an intended beneficiary and claimant under such bond.

SC-21 Post Completion Services. As part of Basic Services, the Architect shall perform a one-year warranty walk-through with the Design-Builder if required by Owner, within one year after Substantial Completion of the Project. The purpose of this walk-through is to ascertain any defects or failures of the Work which may be covered by any equipment or other warranties on the Project. If such defects, deficiencies, or failures are noted, the Architect shall assist the Design-Builder in notifying the appropriate party of the nature of the problem and the applicable warranty. If additional architectural or engineering services are necessary to correct defects resulting from Design-Builder error or defective materials, and not the result of Architect error, such services shall be considered as Additional Services and the Architect shall be compensated accordingly, provided such services have been previously authorized in writing by the Design-Builder.

SC-22 Savings Sharing. To the extent that the Design-Builder shares savings with the Owner on actual costs less than the Guaranteed Maximum Price or other limitation, it is agreed that savings shall be shared between Design-Builder and Architect in the following manner:

Design-Builder: _____________________

Architect: _____________________

SC-23 Limitation of Liability. Design-Builder agrees that to the fullest extent permitted by law, and notwithstanding any other provision of this Agreement, the total liability, in the aggregate, of Architect, its officers, directors, employees, agents, and subconsultants, and any of them, to Owner, the Design-Builder, and anyone claiming by, through, or under them, for any and all claims, damages, losses, liabilities, costs, or damages whatsoever arising out of, resulting from or in any way related to the Project or this Agreement from any cause, including but not limited to the negligence, professional errors or omissions, strict liability, breach of contract, or warranty (express or implied) of Architect, its officers, directors, employees, agents, or consultants or any of them, shall not exceed the Architect's insurance coverage available at the time of payment of such claim or Architect's total fee under this Agreement, whichever is greater. The parties agree that adequate consideration has been given for this limitation.

SC-24 Other Conditions. Add any other modifications below. If there are no others, write "none" below:

____

DESIGN-BUILDER: ARCHITECT:

_______________________   _______________________

By: _______________________   By: _______________________

Name: _______________________   Name: _______________________

Title: _______________________   Title: _______________________

Page 458, after Appendix D add new appendix:

APPENDIX D1

ELECTRONIC DATA AGREEMENT

[Note to user. This agreement should be considered by architects who are asked to furnish their designs to general contractors in electronic version.]

ELECTRONIC DATA AGREEMENT

This Agreement is entered into by _________________________________ (the "Design-Builder") and _________________________________ (the "Architect"). Design-Builder has requested Architect to provide design data in electronic form via electronic mail or on computer disk, consisting of line drawings via computer-aided design (CAD) for the _________________________________ project located at _________________________________ (the "Project"). In consideration of the Architect furnishing such information to Design-Builder, and a limited license to reproduce the same, Design-Builder agrees as follows:

1. Due to the potential that the information set forth on the computer disk can be modified by the Design-Builder or others, unintentionally or otherwise, the Architect shall remove all references to its corporate name, professional seal, and/or involvement from each electronic display.

2. Under no circumstances shall the transfer of ownership of such disk or electronic data, or hard copy thereof, be deemed to be a sale by the Architect of tangible goods, and the Architect provides the data in electronic media merely as an Instrument of Service.

3. The data on the disk represents the information at a particular point in time and Architect shall not be responsible to advise Design-Builder of any changes which may hereafter be made to the Project data or other information contained on the disk or electronic display unless that is a specific requirement stated elsewhere in this Agreement.

4. Architect retains all copyrights to the designs, drawings, information, and works depicted in electronic media as stated in the Design/Builder/Architect Agreement and grants to Design-Builder and Owner a limited license to reproduce such information in connection with Owner's use and occupancy of the Project, and for no other purpose.

5. The Architect specifically disclaims all warranties, expressed or implied, including, but not limited to implied warranties of merchantability and fitness for a particular purpose, with respect to this electronic media and the information contained therein. The Architect shall have no liability with respect to any claim, demand, loss

or damage directly or indirectly arising out of the use of the disk with electronic media contained thereon. Further, Architect shall have no liability for consequential damages, including without limitation damages arising from the alleged negligence of the Architect.

6. The parties understand that the data contained on any disk may be altered, intentionally or otherwise by Design-Builder or others due to occurrences beyond the reasonable control or knowledge of the Architect. These include errors in transcription, machine error, environmental factors, as well as operator error. Design-Builder, as part of the consideration for accepting the delivery of this disk and its use, agrees to indemnify, defend, and hold harmless the Architect, its consultants, and their respective officers, directors, employees, members, and owners from any claims, liabilities, loss, and costs, including, but not limited to, attorneys' fees and cost of defense arising out of changes or modifications to the data and electronic media in the Design-Builder's possession or if released to others by Design-Builder.

7. Architect states that, to the best of its information and belief, at the time of delivery of the electronic media, it accurately represents the information contained on the printed hard copy of Architect's documents. It shall be the Design-Builder's responsibility to verify that the information contained on and displayed by the electronic media disk conforms with the printed hard copy of such drawings and/or specifications provided to Design-Builder. Should there be a discrepancy, Design-Builder agrees to notify the Architect within, but not later than, five (5) days from delivery of the disk/electronic media, whereupon it will be replaced by Architect at no cost. This shall be the sole remedy available for any defect. Failure to so notify the Architect within five (5) days of the delivery of this disk/electronic media shall fully excuse the Architect from any further obligation to Design-Builder hereunder, whatsoever.

Design/Builder: ____

By: _______________________________ Date: _______________________________

Title: _______________________________

Architect: _______________________________

By: _______________________________ Date: _______________________________

Title: _______________________________

STATE DESIGN-BUILD STATUTES (INCLUDING ATTORNEY GENERAL OPINIONS)

KENTUCKY

I. Public Procurement Laws.

Page 549, add the following new statute:

Ky. H.B. 347 (2001). (Signed by Governor 3/20/01).
  Public Procurement Laws.
  AN ACT relating to architectural firms.
  NOTICE: [A> **UPPERCASE TEXT WITHIN THESE SYMBOLS IS ADDED** <A]
  Be it enacted by the General Assembly of the Commonwealth of Kentucky:
  SECTION 1. A NEW SECTION OF KRS CHAPTER 65 IS CREATED TO READ AS FOLLOWS:
  [A> (1) AS USED IN THIS SECTION: <A]
  [A> (A) EMPLOY MEANS TO HIRE, RETAIN, OR OTHERWISE CONTRACT WITH AN INDIVIDUAL OR ENTITY FOR GOODS OR SERVICES; <A]
  [A> (B) LOCAL GOVERNMENT MEANS A CITY, COUNTY, CHARTER COUNTY GOVERNMENT, URBAN-COUNTY GOVERNMENT, CONSOLIDATED LOCAL GOVERNMENT, OR A SPECIAL DISTRICT; <A]
  [A> (C) CONSTRUCTION MANAGER MEANS A PERSON WHO COORDINATES AND COMMUNICATES THE ENTIRE PROJECT PROCESS, CLARIFYING COST AND TIME CONSEQUENCES OF DESIGN DECISIONS AS WELL AS CLARIFYING CONSTRUCTION FEASIBILITY, AND WHO MANAGES THE BIDDING, AWARDING, AND CONSTRUCTION PHASES OF THE PROJECT; AND <A]
  [A> (D) **DESIGN-BUILD** MEANS A SYSTEM OF CONTRACTING UNDER WHICH ONE (1) ENTITY PERFORMS BOTH ARCHITECTURE/ENGINEERING AND CONSTRUCTION UNDER ONE (1) SINGLE CONTRACT. <A]
  [A> (2) A LOCAL GOVERNMENT SHALL NOT EMPLOY THE SAME ENTITY TO PROVIDE BOTH ARCHITECTURAL SERVICES AND CONSTRUCTION MANAGEMENT SERVICES ON THE SAME CAPITAL CONSTRUCTION PROJECT. NO LOCAL GOVERNMENT SHALL KNOWINGLY EMPLOY AN OFFICER, EMPLOYEE, OR AGENT OF, OR AN IMMEDIATE FAMILY MEMBER OF AN OFFICER, EMPLOYEE, OR AGENT OF: <A]
  [A> (A) THE ARCHITECTURAL FIRM THAT PROVIDED THE ARCHITECTURAL SERVICES TO ALSO PROVIDE CONSTRUCTION MANAGEMENT SERVICES FOR THE SAME CAPITAL CONSTRUCTION PROJECT FOR WHICH THE ARCHITECTURAL FIRM PROVIDED ARCHITECTURAL SERVICES; OR <A]

[A> (B) THE CONSTRUCTION MANAGEMENT FIRM THAT PROVIDED THE CONSTRUCTION MANAGEMENT SERVICES TO ALSO PROVIDE ARCHITECTURAL SERVICES FOR THE SAME CAPITAL CONSTRUCTION PROJECT FOR WHICH THE CONSTRUCTION MANAGEMENT FIRM PROVIDED CONSTRUCTION MANAGEMENT SERVICES. <A]

[A> (3) A VIOLATION OF SUBSECTION (2) OF THIS SECTION SHALL SUSPEND THE LOCAL GOVERNMENT FROM RECEIVING ANY FINANCIAL ASSISTANCE FROM THE STATE, OR ANY STATE AGENCY, WITH RESPECT TO THE PROJECT FOR WHICH THE ARCHITECTURAL OR CONSTRUCTION MANAGEMENT FIRM WAS EMPLOYED UNTIL THE MATTER IS RESOLVED. <A]

[A> (4) NOTHING IN THIS SECTION SHALL PROHIBIT A LOCAL GOVERNMENT FROM USING **DESIGN-BUILD** AS A METHOD OF PROVIDING FOR CAPITAL CONSTRUCTION SERVICES. <A]

VIRGINIA

II. Public Procurement Laws.

Page 662, add the following new statute:

Va. S.B. 1049 (2001). (Signed by Governor 3/19/01).

Public Procurement Law.

An Act to amend and reenact Section 33.1-12 of the Code of Virginia, relating to powers and duties of the Commonwealth Transportation Board; *awarding of design-build contracts;* standards for advancing projects from the feasibility stage to the construction stage.

NOTICE: [A> **UPPERCASE TEXT WITHIN THESE SYMBOLS IS ADDED** <A]

Be it enacted by the General Assembly of Virginia:

1. That Section 33.1-12 of the Code of Virginia is amended and reenacted as follows:

Section 33.1-12. General powers and duties of Board; definitions.

The Commonwealth Transportation Board shall be vested with the following powers and shall have the following duties:

(1) Location of routes.—To locate and establish the routes to be followed by the roads comprising systems of state highways between the points designated in the establishment of such systems.

(2) Construction contracts.—[A> (A) <A] To let all contracts for the construction and improvement of the roads comprising systems of state highways and for all activities related to passenger and freight rail and public transportation. [A> (B) THE COMMONWEALTH TRANSPORTATION BOARD MAY AWARD CONTRACTS FOR THE CONSTRUCTION OF TRANSPORTATION PROJECTS ON **A DESIGN-BUILD BASIS.** THE BOARD MAY ANNUALLY AWARD **FIVE DESIGN-BUILD CONTRACTS VALUED NO MORE THAN $20 MILLION.** THE BOARD MAY ALSO AWARD **DESIGN-BUILD CONTRACTS** VALUED MORE THAN $20 MILLION, PROVIDED THAT NO MORE THAN FIVE OF THESE LATTER CONTRACTS ARE IN FORCE AT THE SAME TIME. THESE CONTRACTS MAY BE AWARDED AFTER A WRITTEN DETERMINATION IS MADE BY THE COMMONWEALTH TRANSPORTATION COM-

MISSIONER, PURSUANT TO OBJECTIVE CRITERIA PREVIOUSLY ADOPTED BY THE BOARD REGARDING **THE USE OF DESIGN-BUILD,** THAT DELIVERY OF THE PROJECTS MUST BE EXPEDITED AND THAT IT IS NOT IN THE PUBLIC INTEREST TO COMPLY WITH THE DESIGN AND CONSTRUCTION CONTRACTING PROCEDURES NORMALLY FOLLOWED. SUCH OBJECTIVE CRITERIA WILL INCLUDE REQUIREMENTS FOR PREQUALIFICATION OF CONTRACTORS AND COMPETITIVE BIDDING PROCESSES. THESE CONTRACTS SHALL BE OF SUCH SIZE AND SCOPE TO ENCOURAGE MAXIMUM COMPETITION AND PARTICIPATION BY AGENCY PREQUALIFIED AND OTHERWISE QUALIFIED CONTRACTORS. SUCH DETERMINATION SHALL BE RETAINED FOR PUBLIC INSPECTION IN THE OFFICIAL RECORDS OF THE DEPARTMENT OF TRANSPORTATION AND SHALL INCLUDE A DESCRIPTION OF THE NATURE AND SCOPE OF THE PROJECT AND THE REASONS FOR THE COMMISSIONER'S DETERMINATION THAT AWARDING A **DESIGN-BUILD CONTRACT** WILL BEST SERVE THE PUBLIC INTEREST. THE PROVISIONS OF THIS SECTION SHALL SUPERSEDE CONTRARY PROVISIONS OF SUBDIVISION 2 OF SUBSECTION C OF SECTION 11-41 AND SECTION 11-41.2. <A]

* * *

WASHINGTON

II. Public Procurement Laws.

Page 670, add the following new statute:

Wa. H.B. 1680 (2001). (Passed House 3/9/01; Passed Senate 4/4/01; Enacted 5/9/01)

Public Procurement Law.

AN ACT Relating to **design-build** procedures for public works; adding new sections to chapter 47.20 RCW; adding new sections to chapter 47.60 RCW; and creating a new section.

BE IT ENACTED BY THE LEGISLATURE OF THE STATE OF WASHINGTON:

NEW SECTION. Sec. 1. The legislature finds and declares that a contracting procedure that facilitates construction of transportation facilities in a more timely manner may occasionally be necessary to ensure that construction can proceed simultaneously with the design of the facility. The legislature further finds that **the design-build process** and other alternative project delivery concepts achieve the goals of time savings and avoidance of costly change orders.

The legislature finds and declares that a 2001 audit, conducted by Talbot, Korvola & Warwick, examining the Washington state ferries' capital program resulted in a recommendation for improvements and changes in auto ferry procurement processes. The auditors recommended that auto ferries be procured through use of a modified request for proposals process whereby the prevailing shipbuilder and Washington state ferries **engage in a design and build partnership.** This process promotes ownership of the design by the shipbuilder while using the department of transportation's expertise in ferry design and operations. Alternative processes like **design-build partnerships** can promote innovation and create competitive incentives that increase the likelihood of finishing projects on time and within the budget.

The purpose of this act is to authorize the department's use of a modified request for proposals process for procurement of auto ferries, and to prescribe appropriate requirements and criteria to ensure that contracting procedures for this procurement process serve the public interest.

NEW SECTION. Sec. 2. A new section is added to chapter 47.20 RCW to read as follows:

The department of transportation shall develop a process for awarding competitively bid highway construction contracts for projects over ten million dollars that may be constructed using a ***design-build procedure.*** As used in this section and section 3 of this act, ***"design-build procedure"*** means a method of contracting under which the department of transportation contracts with another party for the party to both ***design and build*** the structures, facilities, and other items specified in the contract.

The process developed by the department must, at a minimum, include the scope of services required ***under the design-build procedure,*** contractor prequalification requirements, criteria for evaluating technical information and project costs, contractor selection criteria, and issue resolution procedures. This section expires April 30, 2008.

NEW SECTION. Sec. 3. A new section is added to chapter 47.20 RCW to read as follows:

The department of transportation may use the ***design-build procedure*** for public works projects over ten million dollars where:

(1) The construction activities are highly specialized and a ***design-build approach*** is critical in developing the construction methodology; or

(2) The projects selected provide opportunity for greater innovation and efficiencies between the designer and the builder; or

(3) Significant savings in project delivery time would be realized.

This section expires April 30, 2008.

NEW SECTION. Sec. 4. A new section is added to chapter 47.60 RCW to read as follows:

(1) The department may purchase new auto ferries through use of a modified request for proposals process whereby the prevailing shipbuilder and the department engage in a ***design and build partnership*** for the design and construction of the auto ferries. The process consists of the three phases described in subsection (2) of this section.

(2) The definitions in this subsection apply throughout sections 5 through 10 of this act.

(a) "Phase one" means the evaluation and selection of proposers to participate in development of technical proposals in phase two.

(b) "Phase two" means the preparation of technical proposals by the selected proposers in consultation with the department.

(c) "Phase three" means the submittal and evaluation of bids, the award of the contract to the successful proposer, and the design and construction of the auto ferries.

NEW SECTION. Sec. 5. A new section is added to chapter 47.60 RCW to read as follows:

To commence the request for proposals process, the department shall publish a notice of its intent once a week for at least two consecutive weeks in at least one trade paper and one other paper, both of general circulation in the state. The notice must contain, but is not limited to, the following information:

(1) The number of auto ferries to be procured, the auto and passenger capacities, the delivery dates, and the estimated price range for the contract;

(2) A statement that a modified request for proposals *design and build partnership* will be used in the procurement process;

(3) A short summary of the requirements for prequalification of proposers including a statement that prequalification is a prerequisite to submittal of a proposal in phase one; and

(4) An address and telephone number that may be used to obtain a prequalification questionnaire and the request for proposals.

NEW SECTION. Sec. 6. A new section is added to chapter 47.60 RCW to read as follows:

Subject to legislative appropriation for the procurement of vessels, the department shall issue a request for proposals to interested parties that must include, at least, the following:

(1) Solicitation of a proposal to participate in a *design and build partnership* with the department to design and construct the auto ferries;

(2) Instructions on the prequalification process and procedures;

(3) A description of the modified request for proposals process. Under this process, the department may modify any component of the request for proposals, including the outline specifications, by addendum at any time before the submittal of bids in phase three;

(4) A description of the *design and build partnership* process to be used for procurement of the vessels;

(5) Outline specifications that provide the requirements for the vessels including, but not limited to, items such as length, beam, displacement, speed, propulsion requirements, capacities for autos and passengers, passenger space characteristics, and crew size. The department will produce notional line drawings depicting hull geometry that will interface with Washington state ferries terminal facilities. Notional lines may be modified in phase two, subject to approval by the department;

(6) Instructions for the development of technical proposals in phase two, and information regarding confidentiality of technical proposals;

(7) The vessel delivery schedule, identification of the port on Puget Sound where delivery must take place, and the location where acceptance trials must be held;

(8) The estimated price range for the contract;

(9) The form and amount of the required bid deposit and contract security;

(10) A copy of the contract that will be signed by the successful proposer;

(11) The date by which proposals in phase one must be received by the department in order to be considered;

(12) A description of information to be submitted in the proposals in phase one concerning each proposer's qualifications, capabilities, and experience;

(13) A statement of the maximum number of proposers that may be selected in phase one for development of technical proposals in phase two;

(14) Criteria that will be used for the phase one selection of proposers to participate in the phase two development of technical proposals;

(15) A description of the process that will be used for the phase three submittal and evaluation of bids, award of the contract, and postaward administrative activities;

(16) A requirement that the contractor comply with all applicable laws, rules, and regulations including but not limited to those pertaining to the environment, worker health and safety, and prevailing wages;

(17) A requirement that the vessels be constructed within the boundaries of the state of Washington except that equipment furnished by the state and components, products, and systems that are standard manufactured items are not subject to the in-state requirement under this subsection. For the purposes of this subsection, "constructed" means the fabrication, by the joining together by welding or fastening of all steel parts from which the total vessel is constructed, including, but not limited to, all shell frames, longitudinals, bulkheads, webs, piping runs, wire ways, and ducting. "Constructed" also means the installation of all components and systems, including, but not limited to, equipment and machinery, castings, electrical, electronics, deck covering, lining, paint, and joiner work required by the contract. "Constructed" also means the interconnection of all equipment, machinery, and services, such as piping, wiring, and ducting; and

(18) A requirement that all warranty work on the vessel must be performed within the boundaries of the state of Washington, insofar as practical.

NEW SECTION. Sec. 7. A new section is added to chapter 47.60 RCW to read as follows:

Phase one of the request for proposals process consists of evaluation and selection of prequalified proposers to participate in subsequent development of technical proposals in phase two, as follows:

(1) The department shall issue a request for proposals to interested parties.

(2) The request for proposals must require that each proposer prequalify for the contract under chapter 468-310 WAC, except that the department may adopt rules for the financial prequalification of proposers for this specific contract only. The department shall modify the financial prequalification rules in chapter 468-310 WAC in order to maximize competition among financially capable and otherwise qualified proposers. In adopting these rules, the department shall consider factors including, without limitation: (a) Shipyard resources in Washington state; (b) the cost to design and construct multiple vessels under a single contract without options; and (c) the sequenced delivery schedule for the vessels.

(3) The department may use some, or all, of the nonfinancial prequalification factors as part of the evaluation factors in phase one to enable the department to select a limited number of best qualified proposers to participate in development of technical proposals in phase two.

(4) The department shall evaluate submitted proposals in accordance with the selection criteria established in the request for proposals. Selection criteria may include, but are not limited to, the following:

    (a) Shipyard facilities;

    (b) Organization components;

    (c) Design capability;

    (d) Build strategy;

(e) Experience and past performance;

(f) Ability to meet vessel delivery dates;

(g) Projected workload; and

(h) Expertise of project team and other key personnel.

(5) Upon concluding its evaluation of proposals, the department shall select the best qualified proposers in accordance with the request for proposals. The selected proposers must participate in development of technical proposals. Selection must be made in accordance with the selection criteria stated in the request for proposals. All proposers must be ranked in order of preference as derived from the same selection criteria.

NEW SECTION. Sec. 8. A new section is added to chapter 47.60 RCW to read as follows:

Phase two of the request for proposals process consists of preparation of technical proposals in consultation with the department, as follows:

(1) The development of technical proposals in compliance with the detailed instructions provided in the request for proposals, including the outline specifications, and any addenda to them. Technical proposals must include the following:

(a) Design and specifications sufficient to fully depict the ferries' characteristics and identify installed equipment;

(b) Drawings showing arrangements of equipment and details necessary for the proposer to develop a firm, fixed price bid;

(c) Project schedule including vessel delivery dates; and

(d) Other appropriate items.

(2) The department shall conduct periodic reviews with each of the selected proposers to consider and critique their designs, drawings, and specifications. These reviews must be held to ensure that technical proposals meet the department's requirements and are responsive to the critiques conducted by the department during the development of technical proposals.

(3) If, as a result of the periodic technical reviews or otherwise, the department determines that it is in the best interests of the department to modify any element of the request for proposals, including the outline specifications, it shall do so by written addenda to the request for proposals.

(4) Proposers must submit final technical proposals for approval that include design, drawings, and specifications at a sufficient level of detail to fully depict the ferries' characteristics and identify installed equipment, and to enable a proposer to deliver a firm, fixed price bid to the department. The department shall reject final technical proposals that modify, fail to conform to, or are not fully responsive to and in compliance with the requirements of the request for proposals, including the outline specifications, as amended by addenda.

NEW SECTION. Sec. 9. A new section is added to chapter 47.60 RCW to read as follows:

Phase three consists of the submittal and evaluation of bids and the award of the contract to the successful proposer for the final design and construction of the auto ferries, as follows:

(1) The department shall request bids for detailed design and construction of the vessels after completion of the review of technical proposals in phase two. The de-

partment will review detailed design drawings in phase three for conformity with the technical proposals submitted in phase two. In no case may the department's review replace the builder's responsibility to deliver a product meeting the phase two technical proposal. The department may only consider bids from selected proposers that have qualified to bid by submitting technical proposals that have been approved by the department.

(2) Each qualified proposer must submit its total bid price for all vessels, including certification that the bid is based upon its approved technical proposal and the request for proposals.

(3) Bids constitute an offer and remain open for ninety days from the date of the bid opening. A deposit in cash, certified check, cashier's check, or surety bond in an amount specified in the request for proposals must accompany each bid and no bid may be considered unless the deposit is enclosed.

(4) The department shall evaluate the submitted bids. Upon completing the bid evaluation, the department may select the responsive and responsible proposer that offers the lowest total bid price for all vessels.

(5) The department may waive informalities in the proposal and bid process, accept a bid from the lowest responsive and responsible proposer, reject any or all bids, republish, and revise or cancel the request for proposals to serve the best interests of the department.

(6) The department may:

(a) Award the contract to the proposer that has been selected as the responsive and responsible proposer that has submitted the lowest total bid price;

(b) If a contract cannot be signed with the apparent successful proposer, award the contract to the next lowest responsive and responsible proposer; or

(c) If necessary, repeat this procedure with each responsive and responsible proposer in order of rank until the list of those proposers has been exhausted.

(7) If the department awards a contract to a proposer under this section, and the proposer fails to enter into the contract and furnish satisfactory contract security as required by chapter 39.08 RCW within twenty days from the date of award, its deposit is forfeited to the state and will be deposited by the state treasurer to the credit of the Puget Sound capital construction account. Upon the execution of a ferry design and construction contract all proposal deposits will be returned.

(8) The department may provide an honorarium to reimburse each unsuccessful phase three proposer for a portion of its technical proposal preparation costs at a preset, fixed amount to be specified in the request for proposals. If the department rejects all bids, the department may provide the honoraria to all phase three proposers that submitted bids.

NEW SECTION. Sec. 10. A new section is added to chapter 47.60 RCW to read as follows:

(1) The department shall immediately notify those proposers that are not selected to participate in development of technical proposals in phase one and those proposers who submit unsuccessful bids in phase three.

(2) The department's decision is conclusive unless an aggrieved proposer files an appeal with the superior court of Thurston county within five days after receiving notice of the department's award decision. The court shall hear any such appeal on the department's administrative record for the project. The court may affirm the deci-

sion of the department, or it may reverse or remand the administrative decision if it determines the action of the department was arbitrary and capricious.

WEST VIRGINIA

Page 671, add the following new statute:

II. Public Procurement Laws.

W.Va. S.B. 298 (2001). Approved by Governor April 23, 2001.

Be it enacted by the Legislature of West Virginia:

That section one, article one, chapter sixty-four of the code of West Virginia, one thousand nine hundred thirty-one, as amended, be amended and reenacted; and that article two of said chapter be amended and reenacted, all to read as follows:

* * * Article 2. Authorization for Department of Administration to Promulgate Legislative Rules.

Section 64-2-1. Department of administration and the auditor.

(a) The legislative rule filed in the state register on the twenty-third day of March, two thousand, authorized under the authority of section six, article twenty-two-a, chapter five of this code, modified by the department of administration to meet the objections of the legislative rule-making review committee and refiled in the state register on the twenty-fourth day of May, two thousand, relating to the department of administration *(rules for selecting design-builders under the design build procurement act, 148 CSR 11),* is authorized with the following amendment:

"On page seven, section eleven, after subsection 11.8 by inserting '11.9. For the purpose of this section, "awarding authority" means the entity having authority to issue and sign the purchase order for the construction or lease-purchase of the project.'"

INDEX

References are to sections or appendixes.

A

Agreement Between Design-Builder and Architect, App. B1

B

Bridging
Canada, 6.01[D]

C

Canada
bridging, 6.01[D]
Canadian Construction Association (CCA), 2.08[A]
Canadian Design-Build Institute (CDBI), 2.08[B]
contract forms
Association of Consulting Engineers of Canada, 18.17[B]
Canadian Construction Documents Committee Forms, 18.17[A]
copyright issues, 16.13
ethical issues, 24.16
compensation from construction professionals, 24.16[E]
compensation from more than one client, 24.16[D]
competition on fees, 24.16[G]
conflicts of interest, 24.16[A]
design competitions restricted, 24.16[F]
free services, 24.16[B]
stipend sharing, 24.16[C]
interference with contract, 24.17
international design-build market, 1.15[D]
licensing requirements
acting as contractor, 12.19[C]
corporations, 12.19[A]
design-led design-build, 12.19[D]
generally, 12.19
partnerships, 12.19[B]
pay applications, 24.03[C]
price lists and tariffs, 24.12[I]
stipends, 14.15[G]
RFQs and RFPs, 14.10
Canadian Design-Build Institute (CDBI), 2.08[B]
bridging, 6.01[D]
RFQs and RFPs, 14.10
Conflicts of Interest
adverse affects of, 24.01[E]
ethical issues, 24.16[A]
Copyright Issues
Canada, 16.13

D

Design-Build Projects
hospitals, 17.14
Design-Build Teaming Agreement
Design-Build Institute of America, 7.06

E

Electronic Data Agreement, App. D1
Ethical Issues in Canada, 24.16
compensation from construction professionals, 24.16[E]
compensation from more than one client, 24.16[D]
competition on fees, 24.16[G]
conflicts of interest, 24.16[A]
design competitions restricted, 24.16[F]
free services, 24.16[B]
stipend sharing, 24.16[C]

F

Fraud
actions based on scope change by owner, 23.12

S

Stipends
payment of in Canada, 14.15[G]